10 Machine Learning Bluep You Should Know for Cybersecurity

Protect your systems and boost your defenses with cutting-edge AI techniques

Rajvardhan Oak

BIRMINGHAM—MUMBAI

10 Machine Learning Blueprints You Should Know for Cybersecurity

Group Product Manager: Ali Abidi
Senior Editor: Rohit Singh
Technical Editor: Kavyashree K S
Copy Editor: Safis Editing
Project Coordinator: Kirti Pisat
Proofreader: Safis Editing
Indexer: Rekha Nair
Production Designer: Arunkumar Govinda Bhat
Developer Relations Marketing Executives: Shifa Ansari and Vinishka Kalra

First published: May 2023

Production reference: 1300523

Published by Packt Publishing Ltd.
Livery Place
35 Livery Street
Birmingham
B3 2PB, UK.

ISBN 978-1-80461-947-6

www.packtpub.com

This book has been a long journey, and I would like to thank my lovely wife, Mrunmayee, for her constant support and unwavering belief in me throughout the process. In a world of chaos, she keeps me sane and inspires me to do great things. This book would not have been possible without her by my side.

– Rajvardhan Oak

Contributors

About the author

Rajvardhan Oak is a cybersecurity expert and researcher passionate about making the Internet a safer place for everyone. His research is focused on using machine learning to solve problems in computer security such as malware, botnets, reputation manipulation, and fake news. He obtained his bachelor's degree from the University of Pune, India, and his master's degree from the University of California, Berkeley. He has been invited to deliver training sessions at summits by the NSF and has served on the program committees of multiple technical conferences. His work has been featured by prominent news outlets such as WIRED magazine and the Daily Mail. In 2022, he received the ISC2 Global Achievement Award for Excellence in Cybersecurity, and in 2023, the honorary Doktor der Akademie from the Akademie für Hochschulbildung, Switzerland. He is based in Seattle and works as an applied scientist in the ads fraud division for Microsoft.

About the reviewers

Dr. Simone Raponi is currently a senior cybersecurity machine learning engineer at Equixely and an ex-machine learning scientist at the NATO Center for Maritime Research and Experimentation. He received both his bachelor's and master's degrees with honor in computer science at the University of Rome, La Sapienza, researching applied security and privacy, and his PhD in computer science and engineering at Hamad Bin Khalifa University in Doha, Qatar, with a focus on cybersecurity and AI. He was awarded the Best PhD in Computer Science and Engineering Award and the Computer Science and Engineering Outstanding Performance Award. His research interest includes cybersecurity, AI, and cyber-threat intelligence.

Duc Haba is a lifelong technologist and researcher. He has been a programmer, enterprise mobility solution architect, AI solution architect, principal, VP, CTO, and CEO. The companies he has worked for range from start-ups and IPOs to enterprise companies.

Duc's career started with Xerox PARC, researching and building expert systems (ruled-based) for copier diagnostic, and skunk works for the USA Department of Defense. Afterward, he joined Oracle, following Viant consulting as a founding member. He dove deep into the entrepreneurial culture in Silicon Valley. There were slightly more failures than successes, but the highlights were Viant and RRKidz. Currently, he is happy working at YML.co as the AI solution architect.

Abhishek Singh is a seasoned professional with almost 15 years of experience in various software engineering roles. Currently, Abhishek serves as a principal software engineer for Azure AI, working on the development of a large-scale distributed AI platform. Previously, Abhishek made significant contributions to cloud and enterprise Security, including founding the Fileless Attack detection capability in Azure Security Center. With a collaborative spirit and a deep-seated understanding of AI, cloud security, and OS internals, Abhishek continuously learns from and contributes to the collective success of various Microsoft products.

Table of Contents

3

Malware Detection Using Transformers and BERT   57

4

Detecting Fake Reviews   81

5

Detecting Deepfakes   107

6

Detecting Machine-Generated Text   127

7

Attributing Authorship and How to Evade It   155

8

Detecting Fake News with Graph Neural Networks   181

9

Attacking Models with Adversarial Machine Learning   205

10

Protecting User Privacy with Differential Privacy   231

11

Protecting User Privacy with Federated Machine Learning   257

12

Breaking into the Sec-ML Industry   283

Preface

Welcome to the wonderful world of cybersecurity and machine learning!

In the 21st century, rapid advancements in technology have brought about incredible opportunities for connectivity, convenience, and innovation. Half a century ago, it would have been hard to believe that you could speak to someone halfway across the world, or that a bot could write stories and poems for you. However, this digital revolution has also introduced new challenges, particularly in the realm of cybersecurity. With each passing day, individuals, businesses, and governments are becoming more reliant on digital systems, making them increasingly vulnerable to cyber threats. As malicious actors grow more sophisticated, it is crucial to develop robust defenses to safeguard our sensitive information, critical infrastructure, and privacy.

Enter machine learning—a powerful branch of artificial intelligence that has emerged as a game-changer in the realm of cybersecurity. Machine learning algorithms have the unique ability to analyze vast amounts of data, identify patterns, and make intelligent predictions. By leveraging this technology, cybersecurity professionals can enhance threat detection, distinguish normal behavior from anomalies, and mitigate risks in real time. Machine learning enables the development of sophisticated intrusion detection systems, fraud detection algorithms, and malware classifiers, empowering defenders to stay one step ahead of cybercriminals. As the digital landscape continues to evolve, the intersection of cybersecurity and machine learning becomes increasingly crucial in safeguarding our digital assets and ensuring a secure and trustworthy future for individuals and organizations alike.

This book presents you with tools and techniques to analyze data and frame a cybersecurity problem as a machine learning task. We will cover multiple forms of cybersecurity, such as the following:

- **System security**, which deals with malware detection, intrusion detection, and adversarial machine learning
- **Application security**, which deals with detecting fake reviews, deepfakes, and fake news
- **Privacy** techniques such as federated machine learning and differential privacy

Throughout the book, I have attempted to use multiple analytical frameworks such as statistical testing, regression, transformers, and graph neural networks. A strong understanding of these will allow you to analyze and solve a cybersecurity problem from multiple approaches.

As new technology is developed, malicious actors come up with new attack strategies. Machine learning is a powerful solution to automatically learn from patterns and detect novel attacks. As a result, there is a high demand in the industry for professionals having expertise at the intersection of cybersecurity and machine learning. This book can help you get started on this wonderful and exciting journey.

Who this book is for

This book is for machine learning practitioners interested in applying their skills to solve cybersecurity issues. Cybersecurity workers looking to leverage ML methods will also find this book useful. An understanding of the fundamental machine learning concepts and beginner-level knowledge of Python programming are needed to grasp the concepts in this book. Whether you're a beginner or an experienced professional, this book offers a unique and valuable learning experience that'll help you develop the skills needed to protect your network and data against the ever-evolving threat landscape.

What this book covers

Chapter 1, On Cybersecurity and Machine Learning, introduces you to the fundamental principles of cybersecurity and how it has evolved, as well as basic concepts in machine learning. It will also discuss the challenges and importance of applying machine learning to the security space.

Chapter 2, Detecting Suspicious Activity, describes the basic cybersecurity problems: detecting intrusions and suspicious behavior that indicates attacks. It will cover statistical and machine learning techniques for anomaly detection.

Chapter 3, Malware Detection Using Transformers and BERT, discusses malware and its variants. A state-of-the-art model, BERT, is used to frame malware detection as an NLP task to build a high-performance classifier with a small amount of malware data. The chapter also covers theoretical details on attention and the transformer model.

Chapter 4, Detecting Fake Reviews, covers techniques for building models for fraudulent review detection. This chapter covers statistical analysis methods such as t-tests to determine which features are statistically different between real and fake reviews. Furthermore, it describes how regression can help model this data and how the results of regression should be interpreted.

Chapter 5, Detecting Deepfakes, discusses deepfake images and videos, which have recently taken the internet by storm. The chapter covers how deepfakes are generated, the social implications they can have, and how machine learning can be used to detect deepfake images and videos.

Chapter 6, Detecting Machine-Generated Text, extends deepfakes into the text domain and covers bot-generated text detection. It first outlines a methodology for generating a custom fake news dataset using GPT, followed by techniques for generating features, and finally, using machine learning to detect text that is bot-generated.

Chapter 7, Attributing Authorship and How to Evade it, talks about the task of authorship attribution, which is important in social media and intellectual privacy domains. The chapter also explores the counter-task – that is, evading authorship attribution – and how that can be achieved to maintain privacy when needed.

Chapter 8, Detecting Fake News with Graph Neural Networks, tackles an important issue in today's world – that of misinformation and fake news. It uses the advanced modeling technique of graph neural networks, explains the theory behind it, and applies it to fake news detection on Twitter.

Chapter 9, Attacking Models with Adversarial Machine Learning, covers security issues related to machine learning models – for example, how a model can be degraded due to data poisoning or how a model can be fooled into giving out an incorrect prediction. You will learn about attack techniques to fool image and text classification models.

Chapter 10, Protecting User Privacy with Differential Privacy, introduces users to differential privacy, a paradigm widely adopted in the technology industry. It also covers the fundamental concepts of privacy, both technical and legal. You will learn how to train fraud detection models in a differentially private manner, and the costs and benefits it brings.

Chapter 11, Protecting User Privacy with Federated Machine Learning, covers a collaborative machine learning technique where multiple entities can co-train a model without having to share any training data. The chapter presents an example of how a deep neural network can be trained in a federated fashion.

Chapter 12, Breaking into the Sec-ML Industry, provides a wealth of resources for you to apply all that you have learned so far and prepare you for interviews in the Sec-ML space. It contains resources for further reading, a question bank for interviews, and blueprints for projects you can build out on your own.

To get the most out of this book

In order to get the most from the book, a firm command of the Python programming language is required. Additionally, some experience with Jupyter Notebook is helpful.

Software/hardware covered in the book	Operating system requirements
PyTorch	Windows, macOS, or Linux
TensorFlow	
Keras	
Scikit-learn	
Miscellaneous – other libraries	

Most of the libraries are available for installation via PyPI. This means that to install a particular library, all you need to do is issue the following command on the terminal:

```
pip install <library_name>
```

If you are using the digital version of this book, we advise you to type the code yourself or access the code from the book's GitHub repository (a link is available in the next section). Doing so will help you avoid any potential errors related to the copying and pasting of code.

Download the example code files

You can download the example code files for this book from GitHub at `https://github.com/PacktPublishing/10-Machine-Learning-Blueprints-You-Should-Know-for-Cybersecurity`. If there's an update to the code, it will be updated in the GitHub repository.

We also have other code bundles from our rich catalog of books and videos available at `https://github.com/PacktPublishing/`. Check them out!

Conventions used

There are a number of text conventions used throughout this book.

`Code in text`: Indicates code words in text, database table names, folder names, filenames, file extensions, pathnames, dummy URLs, user input, and Twitter handles. Here is an example: "Mount the downloaded `WebStorm-10*.dmg` disk image file as another disk in your system."

A block of code is set as follows:

```
import pandas as pd
import numpy as np
import os
from requests import get
}
```

When we wish to draw your attention to a particular part of a code block, the relevant lines or items are set in bold:

```
test_input_fn_a = bert.run_classifier.input_fn_builder(
    features=test_features_a,
    seq_length=MAX_SEQ_LENGTH,
    is_training=False,
    drop_remainder=False)
```

Any command-line input or output is written as follows:

```
pip install -r requirements.txt
```

Bold: Indicates a new term, an important word, or words that you see onscreen. For instance, words in menus or dialog boxes appear in **bold**. Here is an example: "The last column, named **target**, identifies the kind of network attack for every row in the data."

> **Tips or important notes**
> Appear like this.

Get in touch

Feedback from our readers is always welcome.

General feedback: If you have questions about any aspect of this book, email us at customercare@ packtpub.com and mention the book title in the subject of your message.

Errata: Although we have taken every care to ensure the accuracy of our content, mistakes do happen. If you have found a mistake in this book, we would be grateful if you would report this to us. Please visit www.packtpub.com/support/errata and fill in the form.

Piracy: If you come across any illegal copies of our works in any form on the internet, we would be grateful if you would provide us with the location address or website name. Please contact us at copyright@packt.com with a link to the material.

If you are interested in becoming an author: If there is a topic that you have expertise in and you are interested in either writing or contributing to a book, please visit authors.packtpub.com.

Share Your Thoughts

Once you've read *10 Machine Learning Blueprints You Should Know for Cybersecurity*, we'd love to hear your thoughts! Scan the QR code below to go straight to the Amazon review page for this book and share your feedback.

https://packt.link/r/1-804-61947-7

Your review is important to us and the tech community and will help us make sure we're delivering excellent quality content.

Download a free PDF copy of this book

Thanks for purchasing this book!

Do you like to read on the go but are unable to carry your print books everywhere? Is your eBook purchase not compatible with the device of your choice?

Don't worry, now with every Packt book you get a DRM-free PDF version of that book at no cost.

Read anywhere, any place, on any device. Search, copy, and paste code from your favorite technical books directly into your application.

The perks don't stop there, you can get exclusive access to discounts, newsletters, and great free content in your inbox daily

Follow these simple steps to get the benefits:

1. Scan the QR code or visit the link below

https://packt.link/free-ebook/9781804619476

2. Submit your proof of purchase
3. That's it! We'll send your free PDF and other benefits to your email directly

1

On Cybersecurity and Machine Learning

With the dawn of the Information Age, cybersecurity has become a pressing issue in today's society and a skill that is much sought after in industry. Businesses, governments, and individual users are all at risk of security attacks and breaches. The fundamental goal of cybersecurity is to keep users and their data safe. Cybersecurity is a multi-faceted problem, ranging from highly technical domains (cryptography and network attacks) to user-facing domains (detecting hate speech or fraudulent credit card transactions). It helps to prevent sensitive information from being corrupted, avoid financial fraud and losses, and safeguard users and their devices from harmful actors.

A large part of cybersecurity analytics, investigations, and detections are now driven by **machine learning** (**ML**)and "smart" systems. Applying data science and ML to the security space presents a unique set of challenges: the lack of sufficiently labeled data, limitations on powerful models due to the need for explainability, and the need for nearly perfect precision due to high-stakes scenarios. As we progress through the book, you will learn how to handle these critical tasks and apply your ML and data science skills to cybersecurity problems.

In this chapter, we will cover the following main topics:

- The basics of cybersecurity
- An overview of machine learning
- Machine learning – cybersecurity versus other domains

By the end of this chapter, you will have built the foundation for further projects.

The basics of cybersecurity

This book aims to marry two important fields of research: cybersecurity and ML. We will present a brief overview of cybersecurity, how it is defined, what the end goals are, and what problems arise.

Traditional principles of cybersecurity

The fundamental aim of cybersecurity is to keep users and data safe. Traditionally, the goals of cybersecurity were three-fold: **confidentiality, integrity, and availability**, the **CIA** triad.

Let us now examine each of these in depth.

Confidentiality

The confidentiality goal aims to keep data secret from unauthorized parties. Only authorized entities should have access to data.

Confidentiality can be achieved by encrypting data. Encryption is a process where plain-text data is coded into a ciphertext using an encryption key. The ciphertext is not human-readable; a corresponding decryption key is needed to decode the data. Encryption of information being sent over networks prevents attackers from reading the contents, even if they intercept the communication. Encrypting data at rest ensures that adversaries will not be able to read your data, even if the physical data storage is compromised (for example, if an attacker breaks into an office and steals hard drives).

Another approach to ensuring confidentiality is access control. Controlling access to information is the first step in preventing unauthorized exposure or sharing of data (whether intentional or not). Access to data should be granted by following the principle of least privilege; an individual or application must have access to the minimum data that it requires to perform its function. For example, only the finance division in a company should have access to revenue and transaction information. Only system administrators should be able to view network and access logs.

ML can help in detecting abnormal access, suspicious behavior patterns, or traffic from malicious sources, thus ensuring that confidentiality is maintained. For example, suppose an administrator suddenly starts accessing confidential/privileged files outside of the regular pattern for themselves or administrators in general. An ML model may flag it as anomalous and set off alarms so that the administrator can be investigated.

Integrity

The integrity goal ensures that data is trustworthy and tamper-free. If the data can be tampered with, there is no guarantee of its authenticity and accuracy; such data cannot be trusted.

Integrity can be ensured using hashing and checksums. A **checksum** is a numerical value computed by applying a hash function to the data. Even if a single bit of the data were to change, the checksum would change. Digital signatures and certificates also facilitate integrity; once an email or code library has been digitally signed by a user, it cannot be changed; any change requires a new digital signature.

Attackers may wish to compromise the integrity of a system or service for their gain. For example, an attacker may intercept incoming requests to a banking server and change the destination account for any money transfer. Malicious browser extensions may redirect users to a site for gaming traffic and advertising statistics; the original destination URL entered by the user was tampered with. By analyzing the patterns in data coupled with other signals, ML models can detect integrity breaches.

Availability

The first two goals ensure that data is kept secret, tamper-free, and safe from attackers. However, these guarantees are meaningless if authorized users cannot access the data as needed. The availability goal ensures that information is always available to legitimate users of a system.

Attackers may try to compromise availability by executing a **denial-of-service (DoS)** attack, where the target service is bombarded with incoming requests from unauthorized or dummy nodes. For example, an attacker may send millions of dummy queries to a database server. While the attacker may not be able to actually extract any useful information, the server is so overwhelmed that legitimate queries by authorized users are never executed. Alternatively, an attacker may also degrade availability by physically destroying data.

Availability can be guaranteed by implementing redundancy in data and services. Regular backup of data ensures that it is still available even if one copy is destroyed or tampered with. If multiple API endpoints to achieve the same functionality are present, legitimate users can switch to another endpoint, and availability will be preserved. Similar to confidentiality and integrity, pattern analysis and classification algorithms can help detect DoS attacks. An emerging paradigm, graph neural networks, can help detect coordinated bot attacks known as **distributed denial of service (DDoS)**.

Modern cybersecurity – a multi-faceted issue

The CIA triad focuses solely on cybersecurity for data. However, cybersecurity today extends far beyond just data and servers. Data stored in servers becomes information used by organizations, which is transformed into the applications that are ultimately used by humans. Cybersecurity encompasses all these aspects, and there are different problem areas in each aspect.

Figure 1.1 shows how varied and multi-faceted the issue of cybersecurity is and the various elements it encompasses:

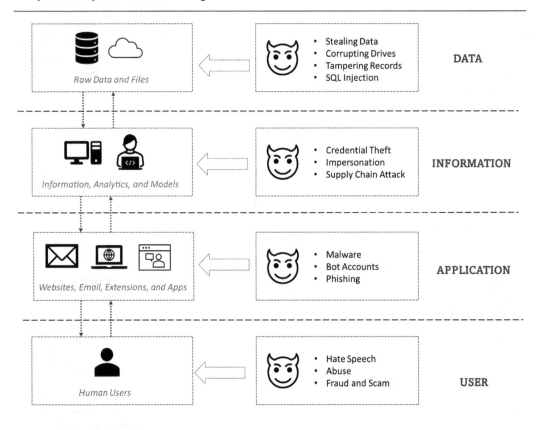

Figure 1.1 – Various problem areas in cybersecurity

In the following sections, we will discuss some common security-related problems and research areas in four broad categories: data security, information security, application security, and user security.

Data security

Data in its raw format is stored on hard drives in offices or in the cloud, which is eventually stored in physical machines in data centers. At the data level, the role of cybersecurity is to keep the data safe. Thus, the focus is on maintaining confidentiality, integrity, and availability (the three goals of the CIA triad) of the data. Cybersecurity problems at this level focus on novel cryptographic schemes, lightweight encryption algorithms, fault tolerance systems, and complying with regulations for data retention.

Information security

Data from data centers and the cloud is transformed into information, which is used by companies to build various services. At the information level, the role of cybersecurity is to ensure that the information is being accessed and handled correctly by employees. Problems at this level focus on network security administration, detecting policy violations and insider threats, and ensuring that there is no inadvertent leakage of private data.

Application security

Information is transformed into a form suitable for consumers using a variety of services. An example of this is information about Facebook users being transformed into a list of top friend recommendations. At the application level, the role of cybersecurity is to ensure that the application cannot be compromised. Problems at this level focus on malware detection, supply chain attack detection, anomaly detection, detecting bot and automated accounts, and flagging phishing emails and malicious URLs.

User security

Finally, applications are used by human end users, and the role of cybersecurity here is to ensure the safety of these users. Problems at this level include detecting hate speech, content moderation, flagging fraudulent transactions, characterizing abusive behavior, and protecting users from digital crimes (identity theft, scams, and extortion). Note that this aspect goes beyond technical elements of cybersecurity and enters into the realm of humanities, law, and the social sciences.

We will now present a case study to explain more clearly how security problems arise at each level (data, information, application, and user).

A study on Twitter

Figure 1.1 shows the threat model in the four broad levels we described. Let us understand this concretely using the example of Twitter. Twitter is a social media networking platform where users can post short opinions, photos, and videos and interact with the posts of others.

At the data level, all of the data (posts, login credentials, and so on) are stored in the raw form. An adversary may try to physically break into data centers or try to gain access to this data using malicious injection queries. At the information level, Twitter itself is using the data to run analytics and train its predictive models. An adversary may try to harvest employee credentials via a phishing email or poison models with corrupt data.

At the application level, the analytics are being transformed into actionable insights consumable by end users: recommendation lists, news feeds, and top-ranking posts. An adversary may create bot accounts that spread misinformation or malicious extensions that redirect users outside of Twitter. Finally, at the user level, end users actually use the app to tweet. Here, an adversary may try to attack users with hate speech or abusive content.

So far, we have discussed cybersecurity and looked at various cybersecurity problems that occur. We will now turn to a related but slightly different topic: privacy.

Privacy

The terms security and privacy are often confused and used interchangeably. However, the goals of security and privacy are very different. While security aims to secure data and systems, privacy refers to individuals having full control over their data. When it comes to privacy, every individual should know the following:

- What data is being collected (location, app usage, web tracking, and health metrics)
- How long it will be retained for (deleted immediately or in a week/month/year)
- Who can access it (advertisers, research organizations, and governments)
- How it can be deleted (how to make a request to the app)

Security and privacy are interrelated. If an attacker hacks into a hospital or medical database (a breach of security), then they may have access to sensitive patient data (a breach of privacy). There are numerous laws around the world, such as the **General Data Protection Regulation (GDPR)** in Europe, that mandate strict security controls in order to ensure user privacy.

Because ML relies on a lot of collected data, there has been a push for privacy-preserving ML. We will be discussing some techniques for this in later chapters.

This completes our discussion of cybersecurity. We started by describing the traditional concepts of security, followed by various cybersecurity problems at multiple levels. We also presented a case study on Twitter that helps put these problems in context. Finally, we looked at privacy, a closely related topic.

Next, we will turn to the second element in the book: ML.

An overview of machine learning

In this section, we will present a brief overview of ML principles and techniques. The traditional computing paradigm defines an algorithm as having three elements: the input, an output, and a process that specifies how to derive the output from the input. For example, in a credit card detection system, a module to flag suspicious transactions may have transaction metadata (location, amount, type) as input and the flag (suspicious or not) as output. The process will define the rule to set the flag based on the input, as shown in *Figure 1.2*:

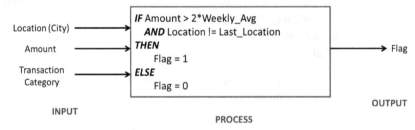

Figure 1.2 – Traditional input-process-output model for fraud detection

ML is a drastic change to the input-process-output philosophy. The traditional approach defined computing as deriving the output by applying the process to the input. In ML, we are given the input and output, and the task is to derive the process that connects the two.

Continuing our analogy of the credit card fraud detection system, we will now be provided with a dataset that has the input features and output flags. Our task is to learn how the flag can be computed based on the input. Once we learn the process, we can generalize it to new data that comes our way in the traditional input-process-output way. Most of the security-related problems we will deal with in this book are *classification* problems. Classification is a task in which data points are assigned discrete, categorical labels (fraud/non-fraud or low-risk/medium-risk/high-risk).

ML is not a one-step process. There are multiple steps involved, such as cleaning and preparing the data in the right form, training and tuning models, deploying them, and monitoring them. In the next section, we will look at the major steps involved in the workflow.

Machine learning workflow

Now, we will see some of the basic steps that go into end-to-end ML. It begins with preprocessing the data to make it fit for use by ML models and ends with monitoring and tuning the models.

Data preprocessing

The first step in any ML experiment is to process the data into a format suitable for ML. Real-world data is often not suitable to be used directly by a model due to two main reasons.

The first reason is variability in format. Data around us is in multiple formats, such as numbers, images, text, audio, and video. All of these need to be converted into numerical representations for a model to consume them. Preprocessing would convert images into matrices, text into word embeddings, and audio streams into time series. Some features of data are discrete; they represent categorical variables. For example, in a dataset about users, the `Country` field would take on string values such as `India`, `Canada`, `China`, and so on. Preprocessing converts such categorical variables into a numerical vector form that can be consumed by a model.

The second reason is noise in real-world data. Measurement inaccuracies, processing errors, and human errors can cause corrupt values to be recorded. For example, a data entry operator might enter your age as 233 instead of 23 by mistakenly pressing the 3 key twice. A web scraper collecting data may face network issues and fail to make a request; some fields in the data will then be missing.

In a nutshell, preprocessing removes noise, handles missing data, and transforms data into a format suitable for consumption by an ML model.

The training phase

After the data has been suitably preprocessed, we train a model to learn from the data. A model expresses the relationship between the preprocessed features and some target variables. For example, a model to detect phishing emails will produce as output a probability that an email is malicious based on the input features we define. A model that classifies malware as 1 of 10 malware families will output a 10-dimensional vector of probability distribution over the 10 classes. In the training phase, the model will learn the parameters of this relationship so that the error over the training data is minimized (that is, it is able to predict the labels for the training data as closely as possible). During the training phase, a subset of the data is reserved as validation data to examine the error of the trained model over unseen data.

One of the problems that a model can face is overfitting; if a model learns parameters too specific to the training data, it cannot generalize well to newer data. This can be diagnosed by comparing the performance of the model on training and validation data; a model that overfits will show decreasing loss or error over the training data but a non-decreasing loss on validation data. The opposite problem to this also exists – the model can suffer from underfitting where it is simply not able to learn from the training data.

The inferencing phase

Once a model has been trained, we want to use it to make predictions for new, unseen data. There is no parameter learning in this phase; we simply plug in the features from the data and inspect the prediction made by the model. Inferencing often happens in real time. For example, every time you use your credit card, the parameters of your transaction (amount, location, and category) are used to run inferencing on a fraud detection model. If the model flags the transaction as suspicious, the transaction is declined.

The maintenance phase

ML models need continuous monitoring and tuning so that their performance does not degrade. A model may become more error-prone in its predictions as time passes because it has been trained on older data. For example, a model to detect misinformation trained in 2019 would have never been exposed to fake news posts about the COVID-19 pandemic. As a result, it never learned the characteristics of COVID-related fake news and may fail to recognize such articles as misinformation. To avoid this, a model has to be retrained on newer data at a regular cadence so that it learns from it.

Additionally, new features may become available that could help improve the performance of the model. We would then need to train the model with the new feature set and check the performance gains. There may be different slices of data in which different classification thresholds show higher accuracy. The model then has to be tuned to use a different threshold in each slice. Monitoring models is an ongoing process, and there are automated tools (called MLOps tools) that offer functionality for continuous monitoring, training, tuning, and alerting of models.

Now, we will look at the fundamental ML paradigms: supervised and unsupervised learning.

Supervised learning

ML has two major flavors: supervised and unsupervised. In **supervised learning**, we have access to labeled data. From the labeled data, we can learn the relation between the data and the labels. The most fundamental example of a supervised learning algorithm is **linear regression.**

Linear regression

Linear regression assumes that the target variable can be expressed as a linear function of the features. We initially start with a linear equation with arbitrary coefficients, and we tune these coefficients as we learn from the data. At a high level and in the simplest form, linear regression works as follows:

1. Let y be the target variable and x_1, x_2, and x_3 be the predictor features. Assuming a linear relationship, our model is $y = a_0 + a_1 x_1 + a_2 x_2 + a_3 x_3$. Here, a_0, a_1, a_2, and a_3 are parameters initially set to random.

2. Consider the first data point from the training set. It will have its own set of predictors (x_1, x_2, and x_3) and the target as ground truth (y). Calculate a predicted value of the target using the equation defined previously; call this predicted value y'.

3. Calculate a *loss* that indicates the error of the prediction. Typically, we use the **ordinary least squares (OLS)** error as the loss function. It is simply the square of the difference between the actual and predicted value of the target: $L = (y - y')^2$.

4. The loss, L, tells us how far off our prediction is from the actual value. We use the loss to modify the parameters of our model. This part is the one where we *learn* from the data.

5. Repeat *steps 2, 3, and 4* over each data point from the training set, and update the parameters as you go. This completes one *epoch* of training.

6. Repeat the preceding steps for multiple epochs.

7. In the end, your parameters will have been tuned to capture the linear relationship between the features and the target.

We will now look at gradient descent, the algorithm that is the heart and soul of linear regression (and many other algorithms).

Gradient descent

The crucial step in the preceding instructions is *step 4*; this is the step where we update the parameters based on the loss. This is typically done using an algorithm called **gradient descent**. Let θ be the parameters of the model; we want to choose the optimal values for θ such that the loss is as small as possible. We calculate the gradient, which is the derivative of the loss with respect to the parameters. We update the parameter θ to its new value θ' based on the gradient as follows:

$$\theta' = \theta - \eta \frac{\delta L}{\delta \theta}$$

The $\frac{\partial L}{\partial \theta}$ gradient represents the slope of the tangent to the loss curve at that particular value of θ. The sign of the gradient will tell us the direction in which we need to change θ in order to reach minima on the loss. We always move in the direction of descending gradient to minimize the loss. For a clearer understanding, carefully observe the loss curve plotted against the parameter in the following chart:

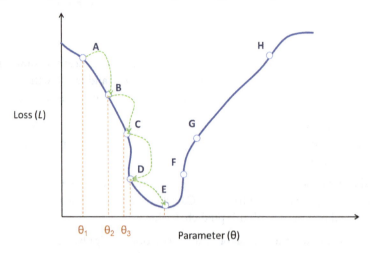

Figure 1.3 – Traversing the loss curve using gradient descent

Suppose because of our random selection, we select the θ_1 parameter. When we calculate the gradient of the curve (the slope of the tangent to the curve at point A), it will be negative. Applying the previous equation of gradient descent, we will have to calculate the gradient and add it to θ_1 to get the new value (say θ_2).

We continue this process until we reach point E, where the gradient is very small (nearly 0); this is the minimum loss we can reach, and even if we were to continue the gradient descent process, the updates would be negligible (because of the very small and near-zero value of the gradient). Had we started at point H instead of A, the gradient we calculated would have been positive. In that case, according to the gradient descent equation, we would have had to decrease θ to reach the minimum loss. Gradient descent ensures that we always move down the curve to reach the minima.

An important element in the equation that we have not discussed so far is η. This parameter is called the **learning rate**. The gradient gives us the direction in which we want to change θ. The learning rate tells us by how much we want to change θ. A smaller value of η means that we are making very small changes and thus taking small steps to reach the minima; it may take a long time to find the optimal values. On the other hand, if we choose a very large value for the learning rate, we may miss the minima.

For example, because of a large learning rate, we might jump directly from point D to F without ever getting to E. At point F, the direction of the gradient changes, and we may jump in the opposite direction back to D. We will keep oscillating between these two points without reaching the minima.

In the linear regression algorithm discussed previously, we performed the gradient descent process for every data point in the training data. This is known as **stochastic gradient descent**. In another, more efficient version of the algorithm, we consider batches of data, aggregate the loss values, and apply gradient descent over the aggregated loss. This is known as **batch gradient descent**.

Gradient descent is at the core of most modern ML algorithms. While we have described it in the context of linear regression, it is simply an optimization algorithm and is widely used in other models such as deep neural networks as well.

Unsupervised learning

In supervised learning, we had ground truth data to learn the relationship between the features and labels. The goal of unsupervised learning is to discover patterns, trends, and relationships within data. Unsupervised learning allows us to make predictions and detect anomalies without having access to labels during training. Let us look at clustering, a popular unsupervised learning problem.

Clustering

As the name suggests, clustering is the process of grouping data into clusters. It allows us to examine how the data points group together, what the common characteristics in those groups are, and what the hidden trends in the data are. There is no *learning* involved in clustering. An ideal clustering is when the intra-cluster similarity is high and inter-cluster similarity is low. This means that points within a cluster are very similar to one another and very different from points in other clusters.

The most fundamental clustering algorithm is called **K-means clustering**. It partitions the data points into k clusters, where k is a parameter that has to be preset. The general process of the K-means algorithm is as follows:

1. Select K points randomly from the training data. These points will be the centroids of our clusters.

2. For each point, calculate the distance to each of the centroids. The distance metric generally used is **Euclidean distance**.

3. Assign each point to one of the K clusters based on the distance. A point will be assigned to the centroid that is closest to it (minimum distance).

4. Each of the K centroids will now have a set of points associated with it; this forms a cluster. For each cluster, calculate the updated cluster centroids as a mean of all the points assigned to that cluster.

5. Repeat *steps 2–4* until one of the following occurs:

    *   The cluster assignment in *step 3* does not change

    *   A fixed number of repetitions of the steps have passed

The core concept behind this algorithm is to optimize the cluster assignment so that the distance of each point to the centroid of the cluster it belongs to is small; that is, clusters should be as tightly knit as possible.

We have looked at supervised and unsupervised learning. A third variant, semi-supervised learning, is the middle ground between the two.

Semi-supervised learning

As the name suggests, semi-supervised learning is the middle ground between supervised and unsupervised learning. Often (and especially in the case of critical security applications), labels are not available or are very few in number. Manual labeling is expensive as it requires both time and expert knowledge. Therefore, we have to train a model from only these limited labels.

Semi-supervised learning techniques are generally based on a self-training approach. First, we train a supervised model based on the small subset of labeled data. We then run inferencing on this model to obtain predicted labels on the unlabeled data. We use our original labels, and high-confidence predicted labels together to train the final model. This process can be repeated for a fixed number of iterations or until we reach a point where the model performance does not change significantly.

Another approach to semi-supervised learning is called **co-training**, where we jointly train two models with different views (feature sets) of the data. We begin by independently training two models based on different feature sets and available labels. We then apply the models to make predictions for the unlabeled data and obtain pseudo labels. We add the high-confidence labels from the first model to the training set of the second and vice versa. We repeat the process of training models and obtain pseudo labels for a fixed number of iterations or until we reach a point where the performance of both models does not change significantly.

Now that we have covered the major paradigms in ML, we can turn to evaluate the performance of models.

Evaluation metrics

So far, we have mentioned model performance in passing. In this section, we will define formal metrics for evaluating the performance of a model. As most of the problems we deal with will be classification-based, we will discuss metrics specific to classification only.

A confusion matrix

Classification involves predicting the label assigned to a particular data point. Consider the case of a fraud detection model. As can be seen in *Figure 1.4*, if the actual and predicted label of a data point are both **Fraud**, then we call that example a **True Positive (TP)**. If both are **Non-Fraud**, we call it a **True Negative (TN)**. If the predicted label is **Fraud**, but the actual label (expected from ground truth) is **Non-Fraud**, then we call it a **False Positive (FP)**. Finally, if the predicted label is **Non-Fraud**, but the actual label (expected from ground truth) is **Fraud**, we call it a **False Negative (FN)**.

Based on the predicted and the actual label of data, we can construct what is called a **confusion matrix**:

PREDICTED CLASS

		Fraud	Non-Fraud
ACTUAL CLASS	*Fraud*	True Positive (**TP**)	False Negative (**FN**)
	Non-Fraud	False Positive (**FP**)	True Negative (**TN**)

Figure 1.4 – Confusion matrix

The confusion matrix provides a quick way to visually inspect the performance of the model. A good model will show the highest true positives and negatives, while the other two will have smaller values. The confusion matrix allows us to conveniently calculate metrics relevant to classification, namely the accuracy, precision, recall, and F-1 score. Let us learn more about these metrics as follows:

- **Accuracy**: This is a measure of how accurate the model predictions are across both classes. Simply put, it is the number of examples for which the model is able to predict the labels correctly, that is, the proportion of true positives and negatives in the data. It can be calculated as follows:

$$Accuracy = \frac{TP + TN}{TP + TN + FP + FN}$$

- **Precision**: This is a measure of the confidence of the model in the positive predictions. In simple terms, it is the proportion of the true positives in all the examples predicted as positive. Precision answers the question: Of all the examples predicted as fraud, how many of them are actually fraud? Precision can be calculated as the following formula:

$$Precision = \frac{TP}{TP + FP}$$

- **Recall**: This is a measure of the completeness of the model in the positive class. Recall answers the question: Of all the examples that were fraud, how many could the model correctly detect as fraud? Recall can be calculated as follows:

$$Recall = \frac{TP}{TP + FN}$$

When a model has high precision, it means that most of the predicted fraud is actually fraud. When a model has a high recall, it means that out of all the fraud in the data, most of it is predicted by the model as fraud. Consider a model that predicts *everything* as fraud. This model has a high recall; as everything is marked as fraud, naturally, all the actual fraud is also being captured, but at the cost of a large number of false positives (low precision). Alternatively, consider a model that predicts *nothing* as fraud. This model has a high precision; as it marks nothing as fraud, there are no false positives, but at the cost of a large number of false negatives (low recall). This trade-off between precision and recall is a classic problem in ML.

- **F-1 Score**: This measure captures the degrees of both precision and recall. High precision comes at the cost of low recall and vice versa. The F-1 score is used to identify the model that has the highest precision and recall together. Mathematically, it is defined as the harmonic mean of the precision and recall, calculated as follows:

$$F1\ Score = \frac{2 \cdot Precision \cdot Recall}{Precision + Recall}$$

- **Receiver Operating Characteristic (ROC) curve**: We have seen that classification models assign a label to a data point. In reality, however, the models compute a probability that the data point belongs to a particular class. The probability is compared with a threshold to determine whether the example belongs to that class or not. Typically, a threshold of 0.5 is used. So, in our example of the fraud detection model, if the model outputs a value greater than 0.5, we will classify the data point as **Fraud**. A smaller value will lead to us classifying it as **Non-Fraud**.

The threshold probability is a parameter, and we can tune this parameter to achieve high precision or recall. If you set a very low threshold, the bar to meet for an example to be fraudulent is very low; for example, at a threshold of 0.1, nearly all examples will be classified as fraud. This means that we will be catching all the fraud out there; we have a high recall.

On the other hand, if you set a very high threshold, the bar to meet will be very high. Not all fraud will be caught, but you can be sure that whatever is marked as fraud is definitely fraud. In other words, you have high precision. This trade-off between precision and recall is captured in the ROC curve, as shown in the following diagram:

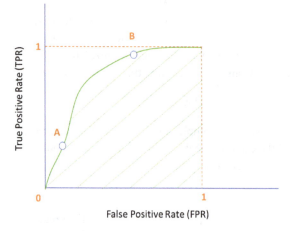

Figure 1.5 – ROC curve

The ROC curve plots the **True Positive Rate (TPR)** against the **False Positive Rate (FPR)** for every value of the threshold from 0 to 1. It allows us to observe the precision we have to tolerate to achieve a certain recall and vice versa. In the ROC plotted in the preceding diagram, points **A** and **B** indicate the true positive and negative rates that we have to tolerate at two different thresholds. The **Area under the**

ROC Curve (AUC), represented by the shaded area, provides an overall measure of the performance of the model across all threshold values. The AUC can be interpreted as the probability that a model scores a random positive example (fraud) higher than a random negative one (non-fraud). When comparing multiple models, we choose the one with the highest value of the AUC.

We have looked at the fundamental concepts behind ML, the underlying workflow, and the evaluation approaches. We will now examine why the application of ML in cybersecurity is different from other fields and the novel challenges it poses.

Machine learning – cybersecurity versus other domains

ML today is applied to a wide variety of domains, some of which are detailed in the following list:

- In sales and marketing, to identify the segment of customers likely to buy a particular product
- In online advertising, for click prediction and to display ads accordingly
- In climate and weather forecasting, to predict trends based on centuries of data
- In recommendation systems, to find the best items (movies, songs, posts, and people) relevant to a user

While every sector imaginable applies ML today, the nuances of it being applied to cybersecurity are different from other fields. In the following subsections, we will see some of the reasons why it is much more challenging to apply ML to the cybersecurity domain than to other domains such as sales or advertising.

High stakes

Security problems often involve making crucial decisions that can impact money, resources, and even life. A fraud detection model that performs well has the potential to save millions of dollars in fraud. A botnet or malware detection model can save critical systems (such as in the military) or sensitive data (such as in healthcare) from being compromised. A model to flag abusive users on social media can potentially save someone's life. Because the stakes are so high, the precision-recall trade-off becomes even more crucial.

In click fraud detection, we have to operate at high precision (or else we would end up marking genuine clicks as fraud), and this comes at the cost of poor recall. Similarly, in abuse detection, we must operate at high recall (we want all abusers to be caught), which comes at the cost of high false positives (low precision). In use cases such as click prediction or targeting, the worst thing that might happen because of a poor model is a few non-converting clicks or mistargeted advertisements. On the other hand, models for security must be thoroughly tuned and diligently monitored as the stakes are much higher than in other domains.

Lack of ground truth

Most well-studied ML methods are supervised in nature. This means that they depend on ground-truth labels to learn the model. In security problems, the ground truth is not always available, unlike in other domains. For example, in a task such as customer targeting or weather forecasting, historical information can tell us whether a customer actually purchased a product or whether it actually rained; this can be used as ground truth.

However, such obvious ground truth is not available in security problems such as fraud or botnet detection. We depend on expert annotation and manual investigations to label data, which is a resource-heavy task. In addition, the labels are based on human knowledge and heuristics (i.e., does this seem fraudulent?) and not on absolute truth such as the ones we described for customer targeting or weather prediction. Due to the lack of high-confidence labels, we often have to rely on unsupervised or semi-supervised techniques for security applications.

The need for user privacy

In recent years, there has been a push to maintain user privacy. As we have discussed previously, with increased privacy comes reduced utility. This proves to be particularly challenging in security tasks. The inability to track users across websites hampers our ability to detect click fraud. A lack of permissions to collect location information will definitely reduce the signals we have access to detect credit card fraud. All of these measures preserve user privacy by avoiding undue tracking and profiling; however, they also degrade the performance of security models, which are actually for the benefit of the user. The fewer signals available to us, the lower the fidelity in the prediction of our models.

The need for interpretability

Most powerful ML models (neural networks, transformers, and graph-based learning models) operate as a black box. The predictions made by the model lack interpretability. Because these models learn higher-order features, it is not possible for a human to understand and explain why a particular example is classified the way it is.

In security applications, however, explainability is important. We need to justify each prediction, at least for the positive class. If a transaction is flagged as fraudulent, the bank or credit card company needs to understand what went into the prediction in order to ascertain the truth. If users are to be blocked or banned, the platform may need adequate justification, more convincing than a model prediction. Classical ML algorithms such as decision trees and logistic regression provide some interpretation based on tree structure and coefficients, respectively. This need for interpretability in models is an obstacle to using state-of-the-art deep learning methods in security.

Data drift

The cybersecurity landscape is an ever-changing one, with new attack strategies coming up every day. The nature of attackers and attack vectors is constantly evolving. As a result, there is also a change in the features that the model expects if data at inference time is significantly different in nature from that which the model was trained on. For example, an entirely new variant of malware may not be detected by a model as no examples of this variant were in the training data. A fake news detection model trained in 2019 may not be able to recognize COVID-19-related misinformation as it was never trained on that data. This data drift makes it challenging to build models that have sustained performance in the wild.

These problems (lack of labels, data drift, and privacy issues) also arise in other applications of ML. However, in the case of systems for cybersecurity, the stakes are high, and the consequences of an incorrect prediction can be devastating. These issues, therefore, are more challenging to handle in cybersecurity.

Summary

This introductory chapter provided a brief overview of cybersecurity and ML. We studied the fundamental goals of traditional cybersecurity and how those goals have now evolved to capture other tasks such as fake news, deep fakes, click spam, and fraud. User privacy, a topic of growing importance in the world, was also introduced. On the ML side, we covered the basics from the ground up: beginning with how ML differs from traditional computing and moving on to the methods, approaches, and common terms used in ML. Finally, we also highlighted the key differences in ML for cybersecurity that make it so much more challenging than other fields. The coming chapters will focus on applying these concepts to designing and implementing ML models for security issues. In the next chapter, we will discuss how to detect anomalies and network attacks using ML.

2

Detecting Suspicious Activity

Many problems in cybersecurity are constructed as anomaly detection tasks, as attacker behavior is generally deviant from good actor behavior. An anomaly is anything that is out of the ordinary—an event that doesn't fit in with normal behavior and hence is considered suspicious. For example, if a person has been consistently using their credit card in Bangalore, a transaction using the same card in Paris might be an anomaly. If a website receives roughly 10,000 visits every day, a day when it receives 2 million visits might be anomalous.

Anomalies are few and rare and indicate behavior that is strange and suspicious. Anomaly detection algorithms are *unsupervised*; we do not have labeled data to train a model. We learn what the normal expected behavior is and flag anything that deviates from it as abnormal. Because labeled data is very rarely available in security-related areas, anomaly detection methods are crucial in identifying attacks, fraud, and intrusions.

In this chapter, we will cover the following main topics:

- Basics of anomaly detection
- Statistical algorithms for intrusion detection
- **Machine learning** (**ML**) algorithms for intrusion detection

By the end of this chapter, you will know how to detect outliers and anomalies using statistical and ML methods.

Technical requirements

All of the implementation in this chapter (and the book too) is using the Python programming language. Most standard computers will allow you to run all of the code without any memory or runtime issues.

There are two options for you to run the code in this book, as follows:

- **Jupyter Notebook**: An interactive code and text notebook with a GUI that will allow you to run code locally. *Real Python* has a very good introductory tutorial on getting started at *Jupyter Notebook: An Introduction* (`https://realpython.com/jupyter-notebook-introduction/`).

- **Google Colab**: This is simply the online version of Jupyter Notebook. You can use the free tier as this is sufficient. Be sure to download any data or files that you create, as they disappear after the runtime is cleaned up.

You will need to install a few libraries that we need for our experiments and analysis. A list of libraries is provided as a text file, and all the libraries can be installed using the `pip` utility with the following command:

```
pip install -r requirements.txt
```

In case you get an error for a particular library not found or installed, you can simply install it with `pip install <library_name>`.

You can find the code files for this chapter on GitHub at `https://github.com/PacktPublishing/10-Machine-Learning-Blueprints-You-Should-Know-for-Cybersecurity/tree/main/Chapter%202`.

Basics of anomaly detection

In this section, we will look at anomaly detection, which forms the foundation for detecting intrusions and suspicious activity.

What is anomaly detection?

The word *anomaly* means *something that deviates from what is standard, normal, or expected*. Anomalies are events or data points that do not fit in with the rest of the data. They represent deviations from the expected trend in data. Anomalies are rare occurrences and, therefore, few in number.

For example, consider a bot or fraud detection model used in a social media website such as Twitter. If we examine the number of follow requests sent to a user per day, we can get a general sense of the trend and plot this data. Let's say that we plotted this data for a month, and ended up with the following trend:

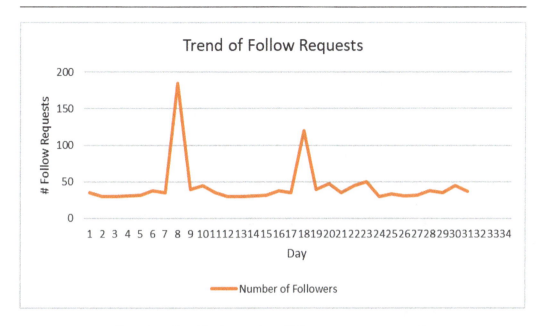

Figure 2.1 – Trend for the number of follow requests over a month

What do you notice? The user seems to have roughly 30-40 follow requests per day. On the 8th and 18th days, however, we see a spike that clearly stands out from the daily trend. These two days are anomalies.

Anomalies can also be visually observed in a two-dimensional space. If we plot all the points in the dataset, the anomalies should stand out as being different from the others. For instance, continuing with the same example, let us say we have a number of features such as the number of messages sent, likes, retweets, and so on by a user. Using all of the features together, we can construct an n-dimensional feature vector for a user. By applying a dimensionality reduction algorithm such as **principal component analysis (PCA)** (at a high level, this algorithm can convert data to lower dimensions and still retain the properties), we can reduce it to two dimensions and plot the data. Say we get a plot as follows, where each point represents a user, and the dimensions represent principal components of the original data. The points colored in red clearly stand out from the rest of the data—these are outliers:

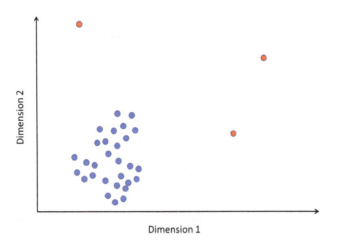

Figure 2.2 – A 2D representation of data with anomalies in red

Note that anomalies do not necessarily represent a malicious event—they simply indicate that the trend deviates from what was normally expected. For example, a user suddenly receiving increased amounts of friend requests is anomalous, but this may have been because they posted some very engaging content. Anomalies, when flagged, must be investigated to determine whether they are malicious or benign.

Anomaly detection is considered an important problem in the field of cybersecurity. Unusual or abnormal events can often indicate security breaches or attacks. Furthermore, anomaly detection does not need labeled data, which is hard to come by in security problems.

Introducing the NSL-KDD dataset

Now that we have introduced what anomaly detection is in sufficient detail, we will look at a real-world dataset that will help us observe and detect anomalies in action.

The data

Before we jump into any algorithms for anomaly detection, let us talk about the dataset we will be using in this chapter. The dataset that is popularly used for anomaly and intrusion detection tasks is the **Network Security Laboratory-Knowledge Discovery in Databases (NSL-KDD)** dataset. This was originally created in 1999 for use in a competition at the 5th *International Conference on Knowledge Discovery and Data Mining (KDD)*. The task in the competition was to develop a network intrusion detector, which is a predictive model that can distinguish between bad connections, called intrusions or attacks, and benign normal connections. This database contains a standard set of data to be audited, which includes a wide variety of intrusions simulated in a military network environment.

Exploratory data analysis (EDA)

This activity consists of a few steps, which we will look at in the next subsections.

Downloading the data

The actual NSL-KDD dataset is fairly large (nearly 4 million records). We will be using a smaller version of the data that is a 10% subset randomly sampled from the whole data. This will make our analysis feasible. You can, of course, experiment by downloading the full data and rerunning our experiments.

First, we import the necessary Python libraries:

```
import pandas as pd
import numpy as np
import os
from requests import get
```

Then, we set the paths for the locations of training and test data, as well as the paths to a label file that holds a header (names of features) for the data:

```
train_data_page = "http://kdd.ics.uci.edu/databases/kddcup99/kddcup.
data_10_percent.gz"
test_data_page = "http://kdd.ics.uci.edu/databases/kddcup99/kddcup.
testdata.unlabeled_10_percent.gz"
labels ="http://kdd.ics.uci.edu/databases/kddcup99/kddcup.names"
datadir = "data"
```

Next, we download the data and column names using the wget command through Python. As these files are zipped (compressed), we have to first extract the contents using the gunzip command. The following Python code snippet does that for us:

```
# Download training data
print("Downloading Training Data")
os.system("wget " + train_data_page)
training_file_name = train_data_page.split("/")[-1].replace(".gz","")
os.system("gunzip " + training_file_name )
with open(training_file_name, "r+") as ff:
  lines = [i.strip().split(",") for i in ff.readlines()]
ff.close()

# Download training column labels
print("Downloading Training Labels")
response = get(labels)
labels = response.text
labels = [i.split(",")[0].split(":") for i in labels.split("\n")]
labels = [i for i in labels if i[0]!='']
final_labels = labels[1::]
```

Finally, we construct a DataFrame from the downloaded streams:

```
data = pd.DataFrame(lines)
labels = final_labels
data.columns = [i[0] for i in labels]+['target']

for i in range(len(labels)):
  if labels[i][1] == ' continuous.':
    data.iloc[:,i] = data.iloc[:,i].astype(float)
```

This completes our step of downloading the data and creating a DataFrame from it. A DataFrame is a tabular data structure that will allow us to manipulate, slice and dice, and filter the data as needed.

Understanding the data

Once the data is downloaded, you can have a look at the DataFrame simply by printing the top five rows:

```
data.head()
```

This should give you an output just like this:

	duration	protocol_type	service	flag	src_bytes	dst_bytes	land	wrong_fragment	urgent	hot	...	dst_host_srv_count	dst_host_same_srv_rate	dst_host_diff_srv_rate	dst_host_same_src_port_rate	dst_host_srv_diff_host_rate	dst_host_serror_rate	dst_host_srv_serror_rate	dst_host_rerror_rate	dst_host_srv_rerror_rate	target
0	0.0	tcp	http	SF	181.0	5450.0	0	0.0	0.0	0.0	...	9.0	1.0	0.0	0.11	0.0	0.0	0.0	0.0	0.0	normal.
1	0.0	tcp	http	SF	239.0	486.0	0	0.0	0.0	0.0	...	19.0	1.0	0.0	0.05	0.0	0.0	0.0	0.0	0.0	normal.
2	0.0	tcp	http	SF	235.0	1337.0	0	0.0	0.0	0.0	...	29.0	1.0	0.0	0.03	0.0	0.0	0.0	0.0	0.0	normal.
3	0.0	tcp	http	SF	219.0	1337.0	0	0.0	0.0	0.0	...	39.0	1.0	0.0	0.03	0.0	0.0	0.0	0.0	0.0	normal.
4	0.0	tcp	http	SF	217.0	2032.0	0	0.0	0.0	0.0	...	49.0	1.0	0.0	0.02	0.0	0.0	0.0	0.0	0.0	normal.

5 rows × 42 columns

Figure 2.3 – Top five rows from the NSL-KDD dataset

As you can see, the top five rows of the data are displayed. The dataset has 42 columns. The last column, named `target`, identifies the kind of network attack for every row in the data. To examine the distribution of network attacks (that is, how many examples of each kind of attack are present), we can run the following statement:

```
data['target'].value_counts()
```

This will list all network attacks and the count (number of rows) for each attack, as follows:

```
smurf.                      280790
neptune.                    107201
normal.                      97278
back.                         2203
satan.                        1589
ipsweep.                      1247
portsweep.                    1040
warezclient.                  1020
teardrop.                      979
pod.                           264
nmap.                          231
guess_passwd.                   53
buffer_overflow.                30
land.                           21
warezmaster.                    20
imap.                           12
rootkit.                        10
loadmodule.                      9
ftp_write.                       8
multihop.                        7
phf.                             4
perl.                            3
spy.                             2
Name: target, dtype: int64
```

Figure 2.4 – Distribution of data by label (attack type)

We can see that there are a variety of attack types present in the data, with the smurf and neptune types accounting for the largest part. Next, we will look at how to model this data using statistical algorithms.

Statistical algorithms for intrusion detection

Now that we have taken a look at the data, let us look at basic statistical algorithms that can help us isolate anomalies and thus identify intrusions.

Univariate outlier detection

In the most basic form of anomaly detection, known as *univariate anomaly detection*, we build a model that considers the trends and detects anomalies based on a single feature at a time. We can build multiple such models, each operating on a single feature of the data.

z-score

This is the most fundamental method to detect outliers and a cornerstone of statistical anomaly detection. It is based on the **central limit theorem (CLT)**, which says that in most observed distributions, data is clustered around the mean. For every data point, we calculate a *z-score* that indicates how far it is from the mean. Because absolute distances would depend on the scale and nature of data, we measure how many standard deviations away from the mean the point falls. If the mean and standard deviation of a feature are μ and σ respectively, the *z-score* for a point *x* is calculated as follows:

$$z = \frac{x - \mu}{\sigma}$$

The value of *z* is the number of standard deviations away from the mean that *x* falls. The CLT says that most data (99%) falls within two standard deviations (on either side) of the mean. Thus, the higher the value of *z*, the higher the chances of the point being an anomaly.

Recall our defining characteristic of anomalies: they are rare and few in number. To simulate this setup, we sample from our dataset. We choose only those rows for which the target is either `normal` or `teardrop`. In this new dataset, the examples labeled `teardrop` are anomalies. We assign a label of 0 to the normal data points and 1 to the anomalous ones:

```
data_resampled = data.loc[data["target"].
isin(["normal.","teardrop."])]

def map_label(target):
  if target == "normal.":
    return 0
  return 1

data_resampled["Label"] = data_resampled ["target"].apply(map_label)
```

As univariate outlier detection operates on only one feature at a time, let us choose `wrong_fragment` as the feature for demonstration. To calculate the *z-score* for every data point, we first calculate the mean and standard deviation of `wrong_fragment`. We then subtract the mean of the entire group from the `wrong_fragment` value in each row and divide it by the standard deviation:

```
mu = data_resampled ['wrong_fragment'].mean()
sigma = data_resampled ['wrong_fragment'].std()
data_resampled["Z"] = (data_resampled ['wrong_fragment'] - mu) / sigma
```

We can plot the distribution of the *z-score* to visually discern the nature of the distribution. The following line of code can generate a density plot:

```
data_resampled["Z"].plot.density()
```

It should give you something like this:

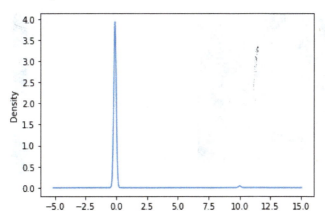

Figure 2.5 – Density plot of z-scores

We see a sharp spike in the density around 0, which indicates that most of the data points have a *z-score* around 0. Also, notice the very small blip around 10; these are the small number of outlier points that have high *z-scores*.

Now, all we have to do is filter out those rows that have a *z-score* of more than 2 or less than -2. We want to assign a label of 1 (predicted anomalies) to these rows, and 0 (predicted normal) to the others:

```
def map_z_to_label(z):
    if z > 2 or z < -2:
        return 1
    return 0
data_resampled["Predicted_Label"] = data_resampled["Z"].apply(map_z_
to_label)
```

Now we have the actual and predicted labels for each row, we can evaluate the performance of our model using the confusion matrix (described earlier in *Chapter 1, On Cybersecurity and Machine Learning*). Fortunately, the scikit-learn package in Python provides a very convenient built-in method that allows us to compute the matrix, and another package called seaborn allows us to quickly plot it. The code is illustrated in the following snippet:

```
from sklearn.metrics import confusion_matrix
from matplotlib import pyplot as plt
import seaborn as sns

confusion = confusion_matrix(data_resampled["Label"],
                data_resampled["Predicted_Label"])
plt.figure(figsize = (10,8))
sns.heatmap(confusion, annot = True, fmt = 'd', cmap="YlGnBu")
```

This will produce a confusion matrix as shown:

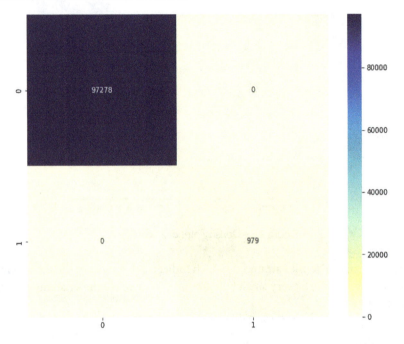

Figure 2.6 – Confusion matrix

Observe the confusion matrix carefully, and compare it with the skeleton confusion matrix from *Chapter 1, On Cybersecurity and Machine Learning*. We can see that our model is able to perform very well; all of the data points have been classified correctly. We have only true positives and negatives, and no false positives or false negatives.

Elliptic envelope

Elliptic envelope is an algorithm to detect anomalies in Gaussian data. At a high level, the algorithm models the data as a high-dimensional Gaussian distribution. The goal is to construct an ellipse covering most of the data—points falling outside the ellipse are anomalies or outliers. Statistical methods such as covariance matrices of features are used to estimate the size and shape of the ellipse.

The concept of elliptic envelope is easier to visualize in a two-dimensional space. A very idealized representation is shown in *Figure 2.7*. The points colored blue are within the boundary of the ellipse and hence considered normal or benign. Points in red fall outside, and hence are anomalies:

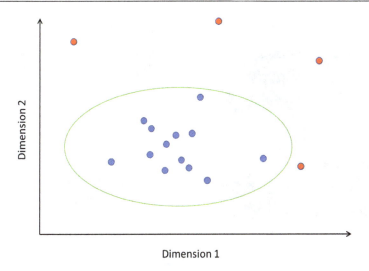

Figure 2.7 – How elliptic envelope detects anomalies

Note that the axes are labeled **Dimension 1** and **Dimension 2**. These dimensions can be features you have extracted from your data; or, in the case of high-dimensional data, they might represent principal component features.

Implementing elliptic envelope as an anomaly detector in Python is straightforward. We will use the resampled data (consisting of only `normal` and `teardrop` data points) and drop the categorical features, as before:

```
from sklearn.covariance import EllipticEnvelope
actual_labels = data4["Label"]
X = data4.drop(["Label", "target",
                "protocol_type", "service",
                "flag"], axis=1)

clf = EllipticEnvelope(contamination=.1,random_state=0)
clf.fit(X)
predicted_labels = clf.predict(X)
```

The implementation of the algorithm is such that it produces -1 if a point is outside the ellipse, and 1 if it is within. To be consistent with our ground truth labeling, we will reassign -1 to 1 and 1 to 0:

```
predicted_labels_rescored =
[1 if pred == -1 else 0 for pred in predicted_labels]
```

Plot the confusion matrix as described previously. The result should be something like this:

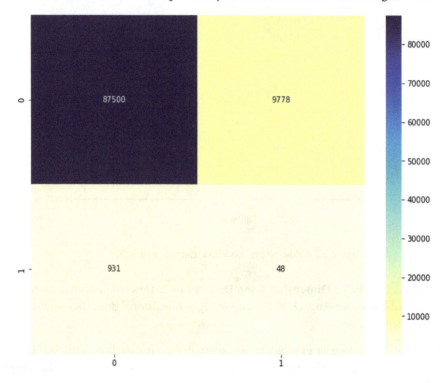

Figure 2.8 – Confusion matrix for elliptic envelope

Note that this model has significant false positives and negatives. While the number may appear small, recall that our data had very few examples of the positive class (labeled 1) to begin with. The confusion matrix tells us that there were 931 false negatives and only 48 true positives. This indicates that the model has extremely low precision and is unable to isolate anomalies properly.

Local outlier factor

Local outlier factor (also known as **LOF**) is a density-based anomaly detection algorithm. It examines points in the local neighborhood of a point to detect whether that point is anomalous. While other algorithms consider a point with respect to the global data distribution, LOF considers only the local neighborhood and determines whether the point fits in. This is particularly useful to identify hidden outliers, which may be part of a cluster of points that is not an anomaly globally. Look at *Figure 2.9*, for instance:

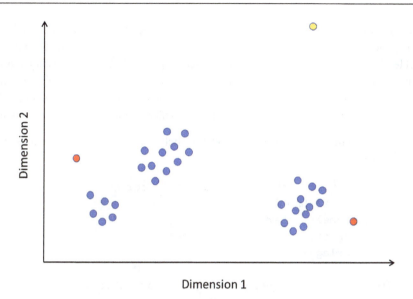

Figure 2.9 – Local outliers

In this figure, the yellow point is clearly an outlier. The points marked in red are not really outliers if you consider the entirety of the data. However, observe the neighborhood of the points; they are far away and stand apart from the local clusters they are in. Therefore, they are anomalies local to the area. LOF can detect such local anomalies in addition to global ones.

In brief, the algorithm works as follows. For every point P, we do the following:

1. Compute the distances from P to every other point in the data. This distance can be computed using a metric called *Manhattan distance*. If (x_1, y_1) and (x_2, y_2) are two distinct points, the Manhattan distance between them is $|x_1 - x_2| + |y_1 - y_2|$. This can be generalized to multiple dimensions.

2. Based on the distance, calculate the K closest points. This is the neighborhood of point P.

3. Calculate the local reachability density, which is nothing but the inverse of the average distance between P and the K closest points. The local reachability density measures how close the neighborhood points are to P. A smaller value of density indicates that P is far away from its neighbors.

4. Finally, calculate the LOF. This is the sum of distances from P to the neighboring points weighted by the sum of densities of the neighborhood points.

5. Based on the LOF, we can determine whether P represents an anomaly in the data or not.

A high LOF value indicates that P is far from its neighbors and the neighbors have high densities (that is, they are close to their neighbors). This means that P is a local outlier in its neighborhood.

A low LOF value indicates that either P is far from the neighbors or that neighbors possibly have low densities themselves. This means that it is not an outlier in the neighborhood.

Note that the performance of our model here depends on the selection of K, the number of neighbors to form the neighborhood. If we set K to be too high, we would basically be looking for outliers at the global dataset level. This would lead to false positives because points that are in a cluster (so not locally anomalous) far away from the high-density regions would also be classified as anomalies. On the other hand, if we set K to be very small, our neighborhood would be very sparse and we would be looking for anomalies with respect to very small regions of points, which would also lead to misclassification.

We can try this out in Python using the built-in off-the-shelf implementation. We will use the same data and features that we used before:

```
from sklearn.neighbors import LocalOutlierFactor
actual_labels = data["Label"]
X = data.drop(["Label", "target",
                "protocol_type", "service",
                "flag"], axis=1)
k = 5
clf = LocalOutlierFactor(n_neighbors=k, contamination=.1)
predicted_labels = clf.fit_predict(X)
```

After rescoring the predicted labels for consistency as described before, we can plot the confusion matrix. You should see something like this:

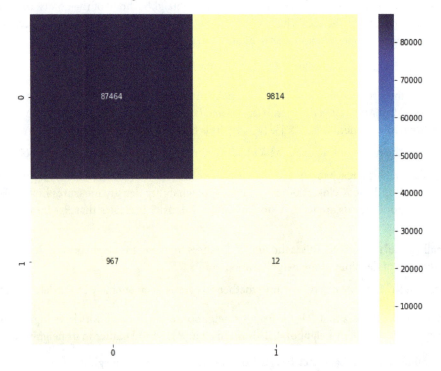

Figure 2.10 – Confusion matrix for LOF with K = 5

Clearly, the model has an extremely high number of false negatives. We can examine how the performance changes by changing the value of *K*. Note that we first picked a very small value. If we rerun the same code with *K* = 250, we get the following confusion matrix:

Figure 2.11 – Confusion matrix for LOF with K = 250

This second model is slightly better than the first. To find the best *K* value, we can try doing this over all possible values of *K*, and observe how our metrics change. We will vary *K* from 100 to 10,000, and for each iteration, we will calculate the accuracy, precision, and recall. We can then plot the trends in metrics with increasing *K* to check which one shows the best performance.

The complete code listing for this is shown next. First, we define empty lists that will hold our measurements (accuracy, precision, and recall) for each value of *K* that we test. We then fit an LOF model and compute the confusion matrix. Recall the definition of a confusion matrix from *Chapter 1, On Cybersecurity and Machine Learning*, and note which entries of the matrix define the true positives, false positives, and false negatives.

Using the matrix, we compute accuracy, precision, and recall, and record them in the arrays. Note the calculation of the precision and recall; we deviate from the formula slightly by adding 1 to the denominator. Why do we do this? In extreme cases, we will have zero true or false positives, and we do not want the denominator to be 0 in order to avoid a division-by-zero error:

```
from sklearn.neighbors import LocalOutlierFactor
actual_labels = data4["Label"]
```

```
X = data4.drop(["Label", "target","protocol_type", "service","flag"],
axis=1)
all_accuracies = []
all_precision = []
all_recall = []
all_k = []
total_num_examples = len(X)
start_k = 100
end_k = 3000

for k in range(start_k, end_k,100):
  print("Checking for k = {}".format(k))

  # Fit a model
  clf = LocalOutlierFactor(n_neighbors=k, contamination=.1)
  predicted_labels = clf.fit_predict(X)
  predicted_labels_rescored = [1 if pred == -1 else 0 for pred in
predicted_labels]
  confusion = confusion_matrix(actual_labels, predicted_labels_
rescored)

  # Calculate metrics
  accuracy = 100 * (confusion[0][0] + confusion[1][1]) / total_num_
examples
  precision = 100 * (confusion[1][1])/(confusion[1][1] + confusion[1]
[0] + 1)
  recall = 100 * (confusion[1][1])/(confusion[1][1] + confusion[0][1]
+ 1)

  # Record metrics
  all_k.append(k)
  all_accuracies.append(accuracy)
  all_precision.append(precision)
  all_recall.append(recall)
```

Once complete, we can plot the three series to show how the value of *K* affects the metrics. We can do this using `matplotlib`, as follows:

```
plt.plot(all_k, all_accuracies, color='green', label = 'Accuracy')
plt.plot(all_k, all_precision, color='blue', label = 'precision')
plt.plot(all_k, all_recall, color='red', label = 'Recall')
plt.show()
```

This is what the output looks like:

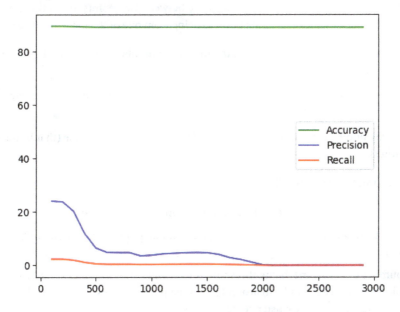

Figure 2.12 – Accuracy, precision, and recall trends with K

We see that while accuracy and recall have remained more or less similar, the value of precision shows a declining trend as we increase *K*.

This completes our discussion of statistical measures and methods for anomaly detection and their application to intrusion detection. In the next section, we will look at a few advanced unsupervised methods for doing the same.

Machine learning algorithms for intrusion detection

This section will cover ML methods such as clustering, autoencoders, SVM, and isolation forests, which can be used for anomaly detection.

Density-based scan (DBSCAN)

In the previous chapter where we introduced **unsupervised ML (UML)**, we studied the concept of clustering via the K-Means clustering algorithm. However, recall that *K* is a hyperparameter that has to be set manually; there is no good way to know the ideal number of clusters in advance. **DBSCAN** is a density-based clustering algorithm that does not need a pre-specified number of clusters.

DBSCAN hinges on two parameters: the minimum number of points required to call a group of points a cluster and ξ (which specifies the minimum distance between two points to call them neighbors). Internally, the algorithm classifies every data point as being from one of the following three categories:

- **Core points** are those that have at least the minimum number of points in the neighborhood defined by a circle of radius ξ

- **Border points**, which are not core points, but in the neighborhood area or cluster of a core point described previously

- An **anomaly point**, which is neither a core point nor reachable from one (that is, not a border point either)

In a nutshell, the process works as follows:

1. Set the values for the parameters, the minimum number of points, and ξ.

2. Choose a starting point (say, A) at random, and find all points at a distance of ξ or less from this point.

3. If the number of points meets the threshold for the minimum number of points, then A is a core point, and a cluster can form around it. All the points at a distance of ξ or less are border points and are added to the cluster centered on A.

4. Repeat these steps for each point. If a point (say, B) added to a cluster also turns out to be a core point, we first form its own cluster and then merge it with the original cluster around A.

Thus, DBSCAN is able to identify high-density regions in the feature space and group points into clusters. A point that does not fall within a cluster is determined to be anomalous or an outlier. DBSCAN offers two powerful advantages over the K-Means clustering discussed earlier:

- We do not need to specify the number of clusters. Oftentimes, when analyzing data, and especially in unsupervised settings, we may not be aware of the number of classes in the data. We, therefore, do not know what the number of clusters should be for anomaly detection. DBSCAN eliminates this problem and forms clusters appropriately based on density.

- DBSCAN is able to find clusters that have arbitrary shapes. Because it is based on the density of regions around the core points, the shape of the clusters need not be circular. K-Means clustering, on the other hand, cannot detect overlapping or arbitrary-shaped clusters.

As an example of the arbitrary shape clusters that DBSCAN forms, consider the following figure. The color of the points represents the ground-truth labels of the classes they belong to. If K-Means is used, circularly shaped clusters are formed that do not separate the two classes. On the other hand, if we use DBSCAN, it is able to properly identify clusters based on density and separate the two classes in a much better manner:

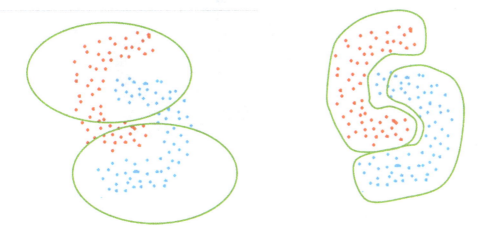

Figure 2.13 – How K-Means (left) and DBSCAN (right) would perform for irregularly shaped clusters

The Python implementation of DBSCAN is as follows:

```
from sklearn.neighbors import DBSCAN
actual_labels = data4["Label"]
X = data4.drop(["Label", "target","protocol_type", "service","flag"],
axis=1)

epsilon = 0.2
minimum_samples = 5
clf = DBSCAN( eps = epsilon,  min_samples = minimum_samples)
predicted_labels = clf.fit_predict(X)
```

After rescoring the labels, we can plot the confusion matrix. You should see something like this:

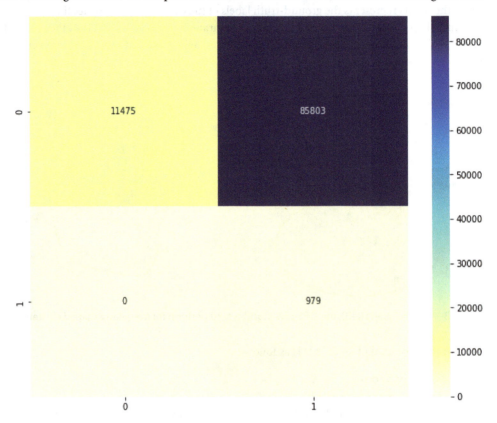

Figure 2.14 – Confusion matrix for DBSCAN

This presents an interesting confusion matrix. We see that all of the outliers are being detected—but along with it, a large number of inliers are also being wrongly classified as outliers. This is a classic case of low precision and high recall.

We can experiment with how this model performs as we vary its two parameters. For example, if we run the same block of code with `minimum_samples = 1000` and `epsilon = 0.8`, we will get the following confusion matrix:

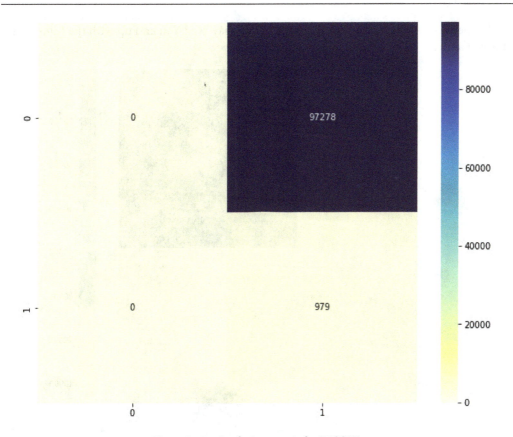

Figure 2.15 – Confusion matrix for DBSCAN

This model is worse than the previous one and is an extreme case of low precision and high recall. Everything is predicted to be an outlier.

What happens if you set `epsilon` to a high value—say, `35`? You end up with the following confusion matrix:

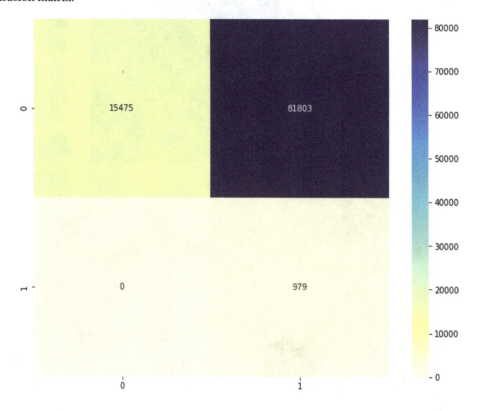

Figure 2.16 – Confusion matrix for improved DBSCAN

Somewhat better than before! You can experiment with other values of the parameters to find out what works best.

One-class SVM

Support vector machine (**SVM**) is an algorithm widely used for classification tasks. **One-class SVM** (**OC-SVM**) is the version used for anomaly detection. However, before we turn to OC-SVM, it would be helpful to have a primer into what SVM is and how it actually works.

Support vector machines

The fundamental goal of SVM is to calculate the optimal decision boundary between two classes, also known as the separating hyperplane. Data points are classified into a category depending on which side of the hyperplane or decision boundary they fall on.

For example, if we consider points in two dimensions, the orange and yellow lines as shown in the following figure are possible hyperplanes. Note that the visualization here is straightforward because we have only two dimensions. For 3D data, the decision boundary will be a plane. For n-dimensional data, the decision boundary will be an n-1-dimensional hyperplane:

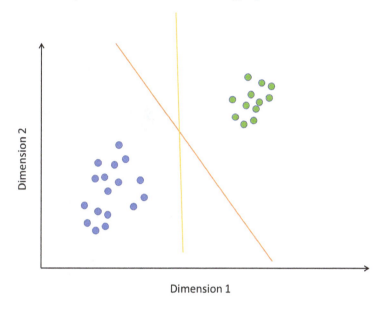

Figure 2.17 – OC-SVM

We can see that in *Figure 2.17*, both the orange and yellow lines are possible hyperplanes. But how does the SVM choose the best hyperplane? It evaluates all possible hyperplanes for the following criteria:

- How well does this hyperplane separate the two classes?

- What is the smallest distance between data points (on either side) and the hyperplane?

Ideally, we want a hyperplane that best separates the two classes and has the largest distance to the data points in either class.

In some cases, points may not be linearly separable. SVM employs the *kernel trick*; using a kernel function, the points are projected to a higher-dimension space. Because of the complex transformations, points that are not linearly separable in their original low-dimensional space may become separable in a higher-dimensional space. *Figure 2.18* (drawn from https://sebastianraschka.com/faq/docs/select_svm_kernels.html) shows an example of a kernel transformation. In the original two-dimensional space, the classes are not linearly separable by a hyperplane. However, when projected into three dimensions, they become linearly separable. The hyperplane calculated in three dimensions is mapped back to the two-dimensional space in order to make predictions:

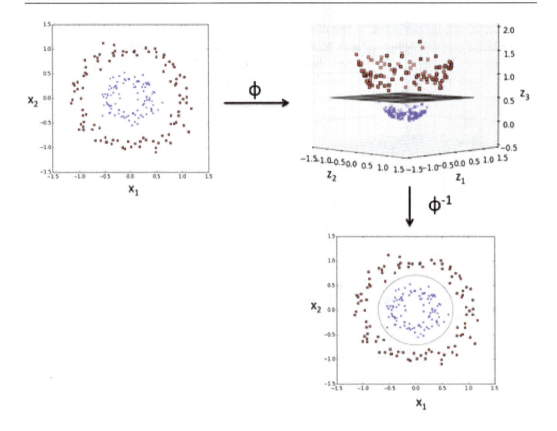

Figure 2.18 – Kernel transformations

SVMs are effective in high-dimensional feature spaces and are also memory efficient since they use a small number of points for training.

The OC-SVM algorithm

Now that we have a sufficient background into what SVM is and how it works, let us discuss OC-SVM and how it can be used for anomaly detection.

OC-SVM has its foundations in the concepts of **Support Vector Data Description** (also known as **SVDD**). While SVM takes a planar approach, the goal in SVDD is to build a *hypersphere* enclosing the data points. Note that as this is anomaly detection, we have no labels. We construct the hypersphere and optimize it to be as small as possible. The hypothesis behind this is that outliers will be removed from the regular points, and will hence fall outside the spherical boundary.

For every point, we calculate the distance to the center of the sphere. If the distance is less than the radius, the point falls inside the sphere and is benign. If it is greater, the point falls outside the sphere and is hence classified as an anomaly.

The OC-SVM algorithm can be implemented in Python as follows:

```
from sklearn import svm

actual_labels = data4["Label"]
X = data4.drop(["Label", "target","protocol_type", "service","flag"],
axis=1)

clf = svm.OneClassSVM(kernel="rbf")
clf.fit(X)
predicted_labels = clf.predict(X)
```

Plotting the confusion matrix, this is what we see:

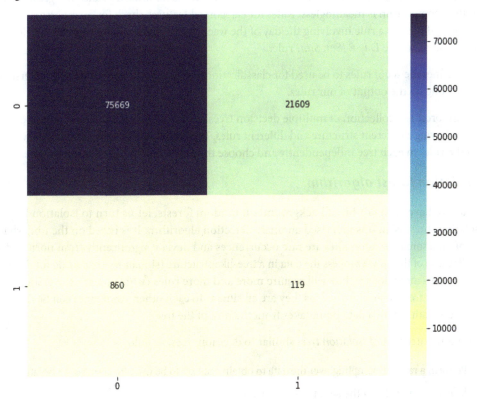

Figure 2.19 – Confusion matrix for OC-SVM

Okay—this seems to be an improvement. While our true positives have decreased, so have our false positives—and by a lot.

Isolation forest

In order to understand what an isolation forest is, it is necessary to have an overview of decision trees and random forests.

Decision trees

A decision tree is an ML algorithm that creates a hierarchy of rules in order to classify a data point. The leaf nodes of a tree represent the labels for classification. All internal nodes (non-leaf) represent rules. For every possible result of the rule, a different child is defined. Rules are such that outputs are generally binary in nature. For continuous features, the rule compares the feature with some value. For example, in our fraud detection decision tree, `Amount > 10,000` is a rule that has outputs as `1` (yes) or `0` (no). In the case of categorical variables, there is no ordering, and so a greater-than or less-than comparison is meaningless. Rules for categorical features check for membership in sets. For example, if there is a rule involving the day of the week, it can check whether the transaction fell on a weekend using the *Day € {Sat, Sun}* rule.

The tree defines the set of rules to be used for classification; we start at the root node and traverse the tree depending on the output of our rules.

A random forest is a collection of multiple decision trees, each trained on a different subset of data and hence having a different structure and different rules. While making a prediction for a data point, we run the rules in each tree independently and choose the predicted label by a majority vote.

The isolation forest algorithm

Now that we have set a fair bit of background on random forests, let us turn to isolation forests. An isolation forest is an unsupervised anomaly detection algorithm. It is based on the underlying definition of anomalies; anomalies are rare occurrences and deviate significantly from normal data points. Because of this, if we process the data in a tree-like structure (similar to what we do for decision trees), the non-anomalous points will require more and more rules (which means traversing more deeply into the tree) to be classified, as they are all similar to each other. Anomalies can be detected based on the path length a data point takes from the root of the tree.

First, we construct a set of *isolation trees* similar to decision trees, as follows:

1. Perform a random sampling over the data to obtain a subset to be used for training an isolation tree.
2. Select a feature from the set of features available.
3. Select a random threshold for the feature (if it is continuous), or a random membership test (if it is categorical).
4. Based on the rule created in *step 3*, data points will be assigned to either the left branch or the right branch of the tree.
5. Repeat *steps 2-4* recursively until each data point is isolated (in a leaf node by itself).

After the isolation trees have been constructed, we have a trained isolation forest. In order to run inferencing, we use an ensemble approach to examine the path length required to isolate a particular point. Points that can be isolated with the fewest number of rules (that is, closer to the root node) are more likely to be anomalies. If we have n training data points, the anomaly score for a point x is calculated as follows:

$$s(x, n) = 2^{-\frac{E(h(x))}{c(n)}}$$

Here, $h(x)$ represents the path length (number of edges of the tree traversed until the point x is isolated). $E(h(x))$, therefore, represents the expected value or mean of all the path lengths across multiple trees in the isolation forest. The constant $c(n)$ is the average path length of an unsuccessful search in a binary search tree; we use it to normalize the expected value of $h(x)$. It is dependent on the number of training examples and can be calculated using the harmonic number $H(n)$ as follows:

$$c(n) = 2H(n - 1) - 2(n - 1)/n$$

The value of $s(x, n)$ is used to determine whether the point x is anomalous or not. Higher values closer to 1 indicate that the points are anomalies, and smaller values indicate that they are normal.

The `scikit-learn` package provides us with an efficient implementation for an isolation forest so that we don't have to do any of the hard work ourselves. We simply fit a model and use it to make predictions. For simplicity of our analysis, let us use only the continuous-valued variables, and ignore the categorical string variables for now.

First, we will use the original DataFrame we constructed and select the columns of interest to us. Note that we must record the labels in a list beforehand since labels cannot be used during model training. We then fit an isolation forest model on our features and use it to make predictions:

```
from sklearn.ensemble import IsolationForest
actual_labels = data["Label"]
X = data.drop(["Label", "target","protocol_type", "service","flag"],
axis=1)
clf = IsolationForest(random_state=0).fit(X)
predicted_labels = clf.predict(X)
```

The `predict` function here calculates the anomaly score and returns a prediction based on that score. A prediction of -1 indicates that the example is determined to be anomalous, and that of 1 indicates that it is not. Recall that our actual labels are in the form of 0 and 1, not -1 and 1. For an apples-to-apples comparison, we will recode the predicted labels, replacing 1 with 0 and -1 with 1:

```
predicted_labels_rescored =
[1 if pred == -1 else 0 for pred in predicted_labels]
```

Now, we can plot a confusion matrix using the actual and predicted labels, in the same way that we did before. On doing so, you should end up with the following plot:

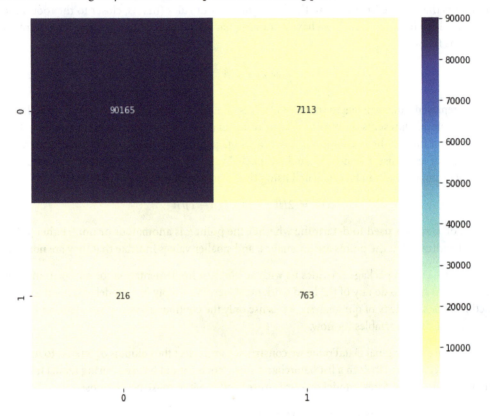

Figure 2.20 – Confusion matrix for isolation forest

We see that while this model predicts the majority of benign classes correctly, it also has a significant chunk of false positives and negatives.

Autoencoders

Autoencoders (**AEs**) are **deep neural networks** (**DNNs**) that can be used for anomaly detection. As this is not an introductory book, we expect you to have some background and a preliminary understanding of **deep learning** (**DL**) and how neural networks work. As a refresher, we will present some basic concepts here. This is not meant to be an exhaustive tutorial on neural networks.

A primer on neural networks

Let us now go through a few basics of neural networks, how they are formed, and how they work.

Neural networks – structure

The fundamental building block of a neural network is a *neuron*. This is a computational unit that takes in multiple numeric inputs and applies a mathematical transformation on it to produce an output. Each input to a neuron has a *weight* associated with it. The neuron first calculates a weighted sum of the inputs and then applies an *activation function* that transforms this sum into an output. Weights represent the parameters of our neural network; training a model means essentially finding the optimum values of the weights such that the classification error is reduced.

A sample neuron is depicted in *Figure 2.21*. It can be generalized to a neuron with any number of inputs, each having its own weight. Here, σ represents the activation function. This determines how the weighted sum of inputs is transformed into an output:

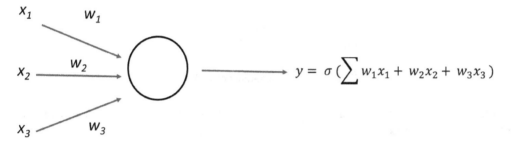

Figure 2.21 – Basic structure of a neuron

A group of neurons together form a *layer* of neurons, and multiple such layers connected together form a neural network. The more the number of layers, the *deeper* the network is said to be. Input data (in the form of features) is fed to the first layer. In the simplest form of neural networks, every neuron in a layer is connected to every neuron in the next layer; this is known as a **fully connected neural network (FCNN)**.

The final layer is the output layer. In the case of binary classification, the output layer has only one neuron with a *sigmoid* activation function. The output of this neuron indicates the probability of the data point belonging to the positive class. If it is a multiclass classification problem, then the final layer contains as many neurons as the number of classes, each with a *softmax* activation. The outputs are normalized such that each one represents the probability of the input belonging to a particular class, and they all add up to 1. All of the layers other than the input and output are not visible from the outside; they are known as *hidden layers*.

Putting all of this together, *Figure 2.22* shows the structure of a neural network:

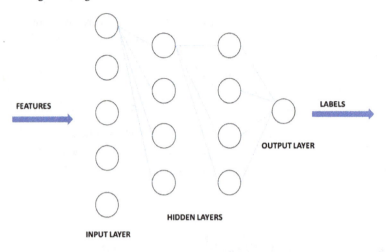

Figure 2.22 – A neural network

Let us take a quick look at four commonly used activation functions: sigmoid, tanh, **rectified linear unit (ReLU)**, and softmax.

Neural networks – activation functions

The *sigmoid* function normalizes a real-valued input into a value between 0 and 1. It is defined as follows:

$$f(z) = \frac{1}{1+e^{-z}}$$

If we plot the function, we end up with a graph as follows:

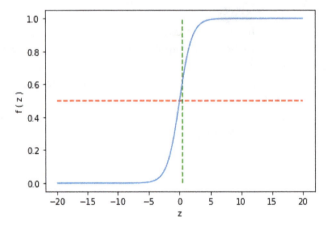

Figure 2.23 – sigmoid activation

As we can see, any number is squashed into a range from 0 to 1, which makes the sigmoid function an ideal candidate for outputting probabilities.

The *hyperbolic tangent* function, also known as the *tanh* function, is defined as follows:

$$\tanh(z) = \frac{2}{1 + e^{-2z}} - 1$$

Plotting this function, we see a graph as follows:

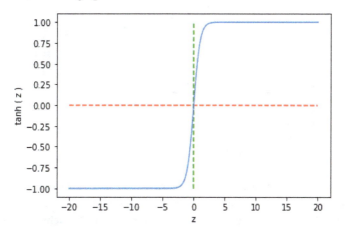

Figure 2.24 – tanh activation

This looks very similar to the sigmoid graph, but note the subtle differences: here, the output of the function ranges from -1 to 1 instead of 0 to 1. Negative numbers get mapped to negative values, and positive numbers get mapped to positive values. As the function is centered on 0 (as opposed to sigmoid centered on 0.5), it provides better mathematical conveniences. Generally, we use *tanh* as an activation function for hidden layers and *sigmoid* for the output layer.

The *ReLU* activation function is defined as follows:

$$ReLU(z) = \begin{cases} \max(0, z); z \geq 0 \\ 0; z < 0 \end{cases}$$

Basically, the output is the same as the input if it is positive; else, it is 0. The graph looks like this:

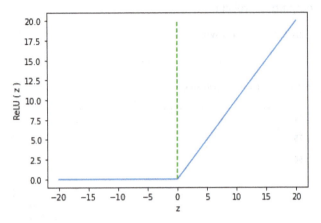

Figure 2.25 – ReLU activation

The *softmax* function can be considered to be a normalized version of sigmoid. While sigmoid operates on a single input, softmax will operate on a vector of inputs and produce a vector as output. Each value in the output vector will be between 0 and 1. If the vector Z is defined as $[z_1, z_2, z_3 ..... z_K]$, softmax is defined as follows:

$$\sigma(z_i) = \frac{e^{z_i}}{\sum_{j=1}^{j=K} e^{z_j}}$$

Basically, we first calculate the sigmoid values for each element in the vector and form a sigmoid vector. Then, we normalize this vector by the total so that each element is between 0 and 1 and they add up to 1 (representing probabilities). This can be better illustrated with a concrete example.

Say that our output vector Z has three elements: Z = [2, 0.9, 0.1]. Then, we have $z_1 = 2$, $z_2 = 0.9$, and $z_3 = 0.1$; we therefore have $e^{z_1} = e^2 = 7.3890$, $e^{z_2} = e^{0.9} = 2.4596$, and $e^{z_3} = e^{0.1} = 1.1051$. Applying the previous equation, the denominator is the sum of these three, which is 10.9537. Now, the output vector is simply the ratio of each element to this summed-up value—that is, $[\frac{7.3890}{10.9537}, \frac{2.4596}{10.9537}, \frac{1.1051}{10.9537}]$, which comes out to be [0.6745, 0.2245, 0.1008]. These values represent probabilities of the input belonging to each class respectively (they do not add up to 1 because of rounding errors).

Now that we have an understanding of activation functions, let us discuss how a neural network actually functions from end to end.

Neural networks – operation

When a data point is passed through a neural network, it undergoes a series of transformations through the neurons in each layer. This phase is known as the forward pass. As the weights are assigned randomly at the beginning, the output at each neuron is different. The final layer will give us the probability of the point belonging to a particular class; we compare this with our ground-truth labels (0 or 1) and calculate a loss. Just as **mean squared error** (MSE) in linear regression is a loss function indicating the error, classification in neural networks uses *binary cross-entropy* and *categorical cross-entropy* as the loss for binary and multiclass classification respectively.

Once the loss is calculated, it is passed back through the network in a process called *backpropagation*. Every weight parameter is adjusted based on how it contributes to the loss. This phase is known as the backward pass. The same gradient descent that we learned before applies here too! Once all the data points have been passed through the network once, we say that a training *epoch* has completed. We continue this process multiple times, with the weights changing and the loss (hopefully) decreasing in each iteration. The training can be stopped either after a fixed number of epochs or if we reach a point where the loss changes only minimally.

Autoencoders – a special class of neural networks

While a standard neural network aims to learn a decision function that will predict the class of an input data point (that is, classification), the goal of autoencoders is to simply *reconstruct* a data point. While training autoencoders, both the input and output that we provide to the autoencoder are the same, which is the feature vector of the data point. The rationale behind it is that because we train only on normal (non-anomalous) data, the neural network will learn how to reconstruct it. On the anomalous data, however, we expect it to fail; remember—the model was never exposed to these data points, and we expect them to be significantly different than the normal ones. As a result, the error in reconstruction is expected to be high for anomalous data points, and we use this error as a metric to determine whether a point is an anomaly.

An autoencoder has two components: an encoder and a decoder. The encoder takes in input data and reduces it to a lower-dimensional space. The output of the encoder can be considered to be a dimensionally reduced version of the input. This output is fed to the decoder, which then projects it into a higher-dimensional subspace (similar to that of the input).

Here is what an autoencoder looks like:

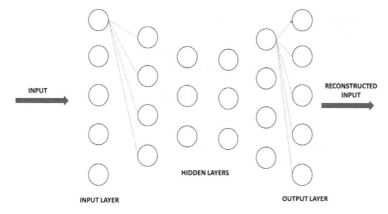

Figure 2.26 – Autoencoder

We will implement this in Python using a framework known as keras. This is a library built on top of the TensorFlow framework that allows us to easily and intuitively design, build, and customize neural network models.

First, we import the necessary libraries and divide our data into training and test sets. Note that while we train the autoencoder only on normal or non-anomalous data, in the real world it is impossible to have a dataset that is 100% clean; there will be some contamination involved. To simulate this, we train on both the normal and abnormal examples, with the abnormal ones being very small in proportion. The stratify parameter ensures that the training and testing data has a similar distribution of labels so as to avoid an imbalanced dataset. The code is illustrated in the following snippet:

```
from tensorflow import keras
from tensorflow.keras import layers
from sklearn.model_selection import train_test_split

X_train, X_test, y_train, y_test = train_test_split(X,
                        actual_labels,
                        test_size=0.33,
                        random_state=42,
                        stratify=actual_labels)

X_train = np.array(X_train, dtype=np.float)
X_test = np.array(X_test, dtype=np.float)
```

We then build our neural network using keras. For each layer, we specify the number of neurons. If our feature vector has N dimensions, the input layer would have to have N layers. Similarly, because this is an autoencoder, the output layer would also have N layers.

We first build the encoder part. We start with an input layer, followed by three fully connected layers of decreasing dimensions. To each layer, we feed the output of the previous layer as input:

```
input = keras.Input(shape=(X_train.shape[1],))
encoded = layers.Dense(30, activation='relu')(input)
encoded = layers.Dense(16, activation='relu')(encoded)
encoded = layers.Dense(8, activation='relu')(encoded)
```

Now, we work our way back to higher dimensions and build the decoder part:

```
decoded = layers.Dense(16, activation='relu')(encoded)
decoded = layers.Dense(30, activation='relu')(decoded)
decoded = layers.Dense(X_train.shape[1], activation='sigmoid')
(encoded)
```

Now, we put it all together to form the autoencoder:

```
autoencoder = keras.Model(input, decoded)
```

We also define the encoder and decoder models separately, to make predictions later:

```
# Encoder
encoder = keras.Model(input_img, encoded)

# Decoder
encoded_input = keras.Input(shape=(encoding_dim,))
decoder_layer = autoencoder.layers[-1]
decoder = keras.Model(encoded_input,decoder_layer(encoded_input))
```

If we want to examine the structure of the autoencoder, we can compile it and print out a summary of the model:

```
autoencoder.compile(optimizer='adam', loss='mse')
autoencoder.summary()
```

You should see which layers are in the network and what their dimensions are as well:

```
Model: "model_13"
_____
 Layer (type)                Output Shape              Param #
=================================================================
 input_14 (InputLayer)       [(None, 38)]              0

 dense_66 (Dense)            (None, 30)                1170

 dense_67 (Dense)            (None, 16)                496

 dense_68 (Dense)            (None, 8)                 136

 dense_69 (Dense)            (None, 16)                144

 dense_70 (Dense)            (None, 30)                510

 dense_71 (Dense)            (None, 38)                1178

=================================================================
Total params: 3,634
Trainable params: 3,634
Non-trainable params: 0
_____
```

Figure 2.27 – Autoencoder model summary

Finally, we actually train the model. Normally, we provide features and the ground truth for training, but in this case, our input and output are the same:

```
autoencoder.fit(X_train, X_train,
                epochs=10,
```

```
                    batch_size=256,
                    shuffle=True)
```

Let's see the result:

```
Total params: 3,634
Trainable params: 3,634
Non-trainable params: 0
```

```
Epoch 1/10
258/258 [==============================] - 1s 1ms/step - loss: 52973192.0000
Epoch 2/10
258/258 [==============================] - 0s 1ms/step - loss: 52973176.0000
Epoch 3/10
258/258 [==============================] - 0s 1ms/step - loss: 52973188.0000
Epoch 4/10
258/258 [==============================] - 0s 1ms/step - loss: 52973176.0000
Epoch 5/10
258/258 [==============================] - 0s 1ms/step - loss: 52973180.0000
Epoch 6/10
258/258 [==============================] - 0s 1ms/step - loss: 52973188.0000
Epoch 7/10
258/258 [==============================] - 0s 1ms/step - loss: 52973148.0000
Epoch 8/10
258/258 [==============================] - 0s 1ms/step - loss: 52973192.0000
Epoch 9/10
258/258 [==============================] - 0s 1ms/step - loss: 52973172.0000
Epoch 10/10
258/258 [==============================] - 0s 1ms/step - loss: 52973172.0000
<keras.callbacks.History at 0x7fbece5e1450>
```

Figure 2.28 – Training phase

Now that the model is fit, we can use it to make predictions. To evaluate this model, we will predict the output for each input data point, and calculate the reconstruction error:

```
from sklearn.metrics import mean_squared_error

predicted = autoencoder.predict(X_test)

errors = [np.linalg.norm(X_test[idx] - k[idx]) for idx in range(X_
test.shape[0])]
```

Now, we need to set a threshold for the reconstruction error, after which we will call data points as anomalies. The easiest way to do this is to leverage the distribution of the reconstruction error. Here, we say that when the error is above the 99th percentile, it represents an anomaly:

```
thresh = np.percentile(errors, 99)
predicted_labels = [1 if errors[idx] > thresh else 0 for idx in
range(X_test.shape[0])]
```

When we generate the confusion matrix, we see something like this:

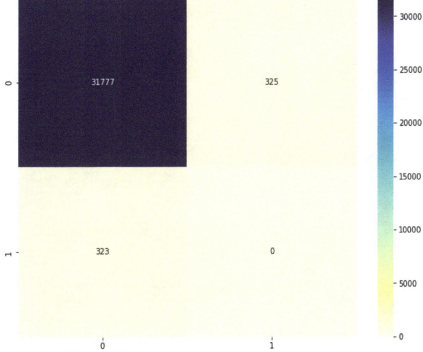

Figure 2.29 – Confusion matrix for Autoencoder

As we can see, this model is a very poor one; there are absolutely no true positives here.

Summary

This chapter delved into the details of anomaly detection. We began by learning what anomalies are and what their occurrence can indicate. Using NSL-KDD, a benchmark dataset, we first explored statistical methods used to detect anomalies, such as the z-score, elliptical envelope, LOF, and DBSCAN. Then, we examined ML methods for the same task, including isolation forests, OC-SVM, and deep autoencoders.

Using the techniques introduced in this chapter, you will be able to examine a dataset and detect anomalous data points. Identifying anomalies is key in many security problems such as intrusion and fraud detection.

In the next chapter, we will learn about malware, and how to detect it using state-of-the-art models known as transformers.

3

Malware Detection Using Transformers and BERT

Malware refers to malicious software applications that run on computers, smartphones, and other devices for nefarious purposes. They execute surreptitiously in the background, and often, users are not even aware that their device is infected with malware. They can be used to steal sensitive user information (such as passwords or banking information) and share it with an adversary, use your device resources for cryptocurrency mining or click fraud, or corrupt your data (such as deleting photos and emails) and ask for a ransom to recover it. In the 21st century, where smartphones are our lifeline, malware can have catastrophic effects. Learning how to identify, detect, and remove malware is an important and emerging problem in cybersecurity.

Because of its ability to identify and learn patterns in behavior, machine learning techniques have been applied to detect malware. This chapter will begin with an overview of malware including its life cycle and operating characteristics. We will then cover an upcoming and state-of-the-art architecture, known as the **transformer**, which is typically used for **natural language processing** (**NLP**) applications. Finally, we will combine the two and show how we can build an extremely high-precision malware classifier using BERT, which is a model built on top of the transformer architecture.

In this chapter, we will cover the following main topics:

- Basics of malware
- Transformers and attention
- Detecting malware with BERT

By the end of this chapter, you will have a better understanding of how malware works. Most importantly, you will be able to apply transformers and BERT to a variety of security-related classification problems based on the concepts we will study here.

Technical requirements

You can find the code files for this chapter on GitHub at `https://github.com/PacktPublishing/10-Machine-Learning-Blueprints-You-Should-Know-for-Cybersecurity/tree/main/Chapter%203`.

Basics of malware

Before we learn about *detecting* malware, let us briefly understand what exactly malware is and how it works.

What is malware?

Malware is simply any *mal*icious soft*ware*. It will install itself on your device (such as a computer, tablet, or smartphone) and operate in the background, often without your knowledge. It is designed to quietly change files on your device, and thus steal or corrupt sensitive information. Malware is generally camouflaged and pretends to be an otherwise innocent application. For example, a browser extension that offers free emojis can actually be malware that is secretly reading your passwords and siphoning them off to a third party.

Devices can be infected by malware in multiple ways. Here are some of the popular vectors attackers exploit to deliver malware to a user device:

- Leveraging the premise of "free" software, such as a cracked version of expensive software such as Adobe Photoshop
- USB devices with the malware installed plugged into the user's computer
- Phishing emails where attackers pretend to be the employer or IT support and ask to download and install a malicious software
- Websites that prompt users to install malicious extensions to continue

An example (`https://www.myantispyware.com/2017/02/20/remove-to-continue-the-work-of-your-browser-you-should-install-the-extension-pop-ups/`) of a website prompting users to install an extension is shown here:

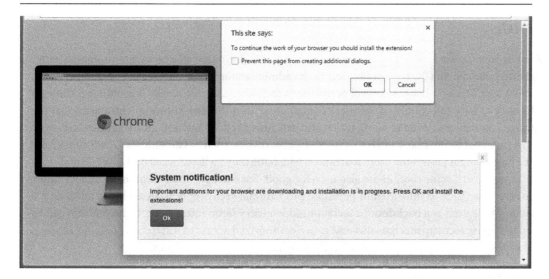

Figure 3.1 – A suspicious website prompting malware downloads

Malware applications come in multiple forms and flavors, each with a different attack strategy. In the next subsection, we will study some popular variants of malware.

Types of malware

Now, let us briefly take a look at the various kinds of malware.

Virus

A virus is a malicious software application that functions in a manner similar to its biological counterpart – an actual virus. A virus program is one that replicates by creating multiple copies of itself and hogs all system resources. Viruses hide in computer files, and once the file is run, they can begin replicating and spreading. Viruses can be boot infectors (which target the operating system directly and install them as part of the booting process) or file infectors (those which are hidden away in executable files, such as free versions of software downloaded from shady websites). Some applications also allow third-party extensions to interface with them. Examples include macros or extensions. Viruses can also run as part of these macros.

Worms

Worms are similar to viruses in operation in terms of their modus operandi, which is to replicate and spread. However, worms are standalone applications; they are not embedded into files like viruses are. While viruses require users to execute the file (such as a .exe file or a macro), worms are more dangerous because they can execute by themselves. Once they infect a computer, they can automatically replicate across the entire network. Worms generally crash the device or overload the network by increasing resource usage.

Rootkits

Rootkits are malware applications that work with the goal of getting the attacker complete administrative rights to your system (the term **root** refers to the administrator or master user in operating systems, and is a user account that can control permissions and other user accounts). A rootkit can allow an adversary to have full control of the user's computer without the user knowing. This means that the attacker can read and write to all files, execute malicious applications, and lock legitimate users out of the system. The attacker can also execute illegal activities, such as launching a DDoS attack, and avoid being caught (as it is the user's machine to which the crime will be traced back). While malicious for the most part, some rootkits are also used for good. For example, if a system that contains highly sensitive data (such as information pertaining to national security) is accessed by an adversary, a rootkit can be used as a **backdoor** (a secret or hidden entry point into a system, and one that can be used to bypass security mechanisms and gain unauthorized access to a system or data) to access it and wipe it off to prevent the information from falling into the wrong hands.

Ransomware

Ransomware are malware applications that block user access to data. Ransomware may encrypt the data so that users are powerless to do anything with their devices. Attackers ask for a *ransom* to be paid, and threaten that the data will be deleted forever or published on the Internet if they do not receive the ransom. There is no guarantee that the attacker will actually hold up their end of the bargain once the ransom is paid. Ransomware attacks have been on the rise, and the emergence of cryptocurrency (such as BTC) has made it possible for attackers to receive and spend money pseudo-anonymously.

Keyloggers

A keylogger is an application that records the keyboard activities of a user. All information typed is logged as keystrokes and siphoned off to an attacker. The attacker can extract information such as usernames, passwords, credit card numbers, and secure PINs. As keyloggers do not cause any visible harm (such as deleting or locking files), they are hard to detect from a user's perspective. They sit quietly in the background and ship your keystroke information to the attacker.

Now that we have explored the different kinds of malware, let us turn to how malware detectors work.

Malware detection

As the prevalence of malware grows, so does the need for detecting it. Routine system scans and analysis by malware detection algorithms can help users stay safe and keep their systems clean.

Malware detection methods

Malware detection can be divided broadly into three main categories: signature-based, behavioral-based, and heuristic methods. In this section, we will look at what these methods are in short and also discuss techniques for analysis.

Signature-based methods

These methods aim to detect malware by storing a database of known malware examples. All applications are checked against this database to identify whether they are malicious. The algorithm examines each application and calculates a signature using a hash function. In computer security, the hash of a file can be treated as its unique identity. It is nearly impossible to have two files with the same hash unless they are identical. Therefore, this method works really well in detecting known malware. While the simplicity of this technique is unmatched, it is easily thwarted; a change of even a single bit in the executable file will cause the hash to be completely different, and undetectable by its signature.

Behavioral-based methods

These methods aim to detect malware by looking for evidence of certain malicious activity. Signature-based methods detect malware based on what the application says, but behavioral methods detect it based on what the application does. It can collect a variety of features from the behavior of the application, such as:

- How many GET requests did the app make?

- How many suspicious URLs did it connect to?

- Does the application have access to file storage?

- How many distinct IP addresses did the application contact in the past seven days?

Using these features, common-sense rules can be built to flag malicious behavior. Past examples of known malware are also studied in detail to identify strategies that can be checked for. Behavioral methods are more robust against evasion, as an adversary will have to explicitly change the behavior of an app to avoid detection.

Heuristic methods

These are the most powerful methods known to us. Rather than look for a specific behavior, they use data mining and machine learning models to learn what malicious applications look like. These methods leverage API calls, OpCode Sequences, call graphs, and other features from the application and train a classification model. Neural networks and Random Forests have been shown to achieve a high-accuracy and high-precision classifier for malware. Heuristic methods are even more robust than behavioral methods, as changing specific parameters may not necessarily fool the model.

Malware analysis

In the previous section, we discussed malware detection methods. Once a potential malware application has been flagged, it needs to be examined to identify its behavior, method of spreading, origin, and any potential impact. Researchers often dissect malware as it can provide insights into the skills and tactics available to an adversary. This process of examining a malware file in detail is known as malware analysis. There are two methods for malware analysis: static and dynamic.

Static analysis

This method examines the malware file as a whole by collecting information about the application without actually running it. The hash of the application is checked against known malware samples. The executable file is decompiled and the code is analyzed in detail; this provides a deep insight into what the goal of the malware was and what the adversary was looking for. Common patterns in the code may also indicate the origin or developer of the malware. Any strategies found can now be used to develop stronger detection mechanisms.

Dynamic analysis

Dynamic analysis involves studying malware by actually executing it. A protected sandbox environment is created, and the malware is allowed to execute in it. This allows researchers the opportunity to look at the malware in action. Some behavior may not be obvious in the code or may dynamically evolve at runtime. Such behavior can be observed when the malware is actually running. Moreover, allowing the application to run allows you to collect API call sequences and other behavioral features, which can be used for heuristic methods.

It is important to note that handling malware can be a dangerous task. Inadvertently running it may cause the virus or Trojan to take control of your system. There are several commercial tools that facilitate malware analysis in a secure way. In later sections, we will be using files generated by one such tool.

Transformers and attention

Transformers are an architecture taking the machine learning world by storm, especially in the fields of natural language processing. An improvement over classical **recurrent neural networks** (**RNN**) for sequence modeling, transformers work on the principle of attention. In this section, we will discuss the attention mechanism, transformers, and the BERT architecture.

Understanding attention

We will now take a look at *attention*, a recent deep learning paradigm that has made great advances in the world of natural language processing.

Sequence-to-sequence models

Most natural language tasks rely heavily on sequence-to-sequence models. While traditional methods are used for classifying a particular data point, sequence-to-sequence architectures map sequences in one domain to sequences in another. An excellent example of this is language translation. An automatic machine translator will take in sequences of tokens (sentences and words) from the source language and map them to other sentences in the target language.

A sequence-to-sequence model generally has two components: the encoder and the decoder. The encoder takes in the source sequences as input and maps them to an intermediate vector known as a **context vector**, or embedding. The decoder takes in the embedding and maps it to sequences in the target domain. The entire model is trained end to end instead of encoders and decoders being trained separately as shown in *Figure 3.2*:

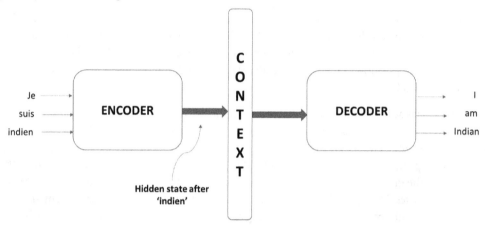

Figure 3.2 – Traditional sequence-to-sequence architecture

The encoder and decoder are typically RNNs, which maintain an internal state (hidden state) that has some memory of past inputs. In a traditional sequence-to-sequence model, the context vector would simply be a high-dimensional representation of the input sentence in a vector space. In *Figure 3.2*, the words from the French sentence are passed one by one to the model (one every time step). The encoder contains an RNN that maintains some memory at every time step. After the final time step, the hidden state of the RNN becomes the context vector.

This is similar to autoencoders discussed previously, except for one major difference: in sequence-to-sequence models, the input and output sequences can be of different lengths. This is often the case in language translation. For example, the French sentence "Ca va?" translates to "How are you?" in English. Sequence-to-sequence models are powerful because they learn the relationships and order between tokens and map them to tokens in the target language.

Attention

The key challenge in the standard encoder/decoder architecture is the bottleneck created by the context vector. Being a fixed-size vector, there is a limitation on how much information can be compressed into it. As a result, it cannot retain information from longer sequences and cannot capture information from multiple timesteps that the RNN encoder goes through. RNNs have a tendency to forget the information they learn. In longer sequences, they will remember the later parts of the sequence and start forgetting earlier ones.

The attention mechanism aims to solve the problem of long-term dependencies and allow the decoder to access as much information as it needs in order to decode the sequence properly. Via attention, the decoder focuses only on the relevant parts of the input sequence in order to produce the output sequence. The model examines multiple time steps from the encoder and "pays attention" to only the ones that it deems to be important.

Concretely speaking, while a traditional sequence-to-sequence model would just pass the last hidden state of the RNN to the decoder, an attention model will pass all of the hidden states. For example, in an English-French translation model, input sentences are English sentences. A hidden state will be created at every word position as the RNN encoder steps through the sequence, and all of these will be passed to the decoder.

The decoder now has access to the context vector at each time step in the input. While decoding, it will try to focus on the parts of the input that are meaningful to decoding this time step. It will examine the encoder's hidden states (remember, all of the hidden states have been passed) and score each hidden state. The score represents the relevance of that hidden state to the current word being decoded; the higher the score, the greater the relevance. The score for each state is normalized using a `softmax` function overall scores. Finally, each hidden state is multiplied by the `softmax` transformed score. Hidden states with high scores (relevant to decoding at this time step) are amplified in value, whereas the ones with low scores are diminished. Using the values of these vectors, the decoded output word can be produced.

Attention in action

We discussed that the attention decoder is able to selectively pay attention to the relevant words in the source sequence. To demonstrate that the model does not mindlessly do a word-by-word translation, we show an example here from the paper that first presented the idea of attention (`https://arxiv.org/abs/1409.0473`).

Consider the problem of translating French sentences into English. This is the perfect domain in which to demonstrate attention. The French language has a peculiar ordering of the parts of speech (adverbs, adjectives, and nouns) that is different from English. If a model is doing a word-by-word translation without attention, the translated output would be grammatically incorrect.

Here is a confusion matrix that demonstrates the attention that the model paid to specific tokens in the input to generate specific tokens in the output. The brighter the color, the stronger the attention:

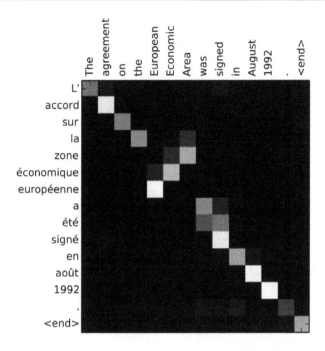

Figure 3.3 – Confusion matrix denoting attention

In French, the adjective is generally placed after the noun. So the "European Economic Zone" becomes the "Zone économique européenne." Looking at the confusion matrix, note how the model has paid attention to the correct word pairs, irrespective of their order. If a model was simply mapping words, the sentence would have been translated from the French version to "Area Economic European." The confusion matrix shows that irrespective of the order, the model knew what words to pay attention to while decoding certain time steps.

This is the fundamental concept of attention. The actual mechanisms (how the hidden states are scored, and how the feature vector is constructed) are out of scope here. However, those of you who are interested can refer to the foundational paper behind attention for a detailed description of the mechanism.

Understanding transformers

In the previous section, we discussed the attention mechanism and how it helps sequence-to-sequence applications such as neural machine translation. Now, we will look at the transformer: an architecture that leverages attention in multiple forms and stages to get the best out of it.

The fundamental architecture of the transformer model is reproduced in *Figure 3.4* from a 2017 paper (`https://arxiv.org/pdf/1706.03762.pdf`) for convenience:

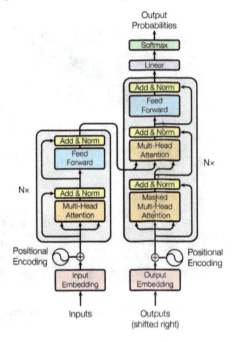

Figure 3.4 – Transformer architecture

The model has two components: the encoder (depicted by the blocks on the left) and the decoder (depicted by the blocks on the right). The goal of the blocks on the left is to take in the input sequence and transform it into the context vectors that are fed to the decoder. The decoder blocks on the right receive the output of the encoder along with the output of the decoder at the previous time step to generate the output sequence. Let us now look at the encoder and decoder blocks in more detail.

The encoder

The encoder consists of two modules: a multi-headed attention module and a fully connected feed-forward neural network. The multi-headed attention module will apply a technique called self-attention, which allows the model to associate each word in the input with other words. For example, consider the sentence, *I could not drink the soup because it was too hot*. Here, the word *it* in the latter half of the sentence refers to the word *soup* in the first half. Self-attention would be able to discover such relationships. As the name suggests, multi-headed attention includes multiple blocks (or heads) for attention. The expectation is that each head will learn a different relationship. The multi-headed attention module calculates the attention weights for the input sequence and generates a vector as output that indicates how each word in the sequence should *pay attention* to the others.

The output of the attention module is added back to the input and then passed through a normalization layer. Normalization helps control the range of parameters and keeps the model stable. The normalized output is passed to the second module, which is a feed-forward neural network. This can be any neural network, but generally speaking, it consists of multiple fully connected layers with a ReLU activation. The addition and normalization process repeats once again, and the final encoder output is produced.

The decoder

The role of the decoder is to take in the encoder's output and produce an output sequence. Note that the general modules remain the same, but the structure differs slightly.

The decoder has two attention modules. One of them takes in the embedding of the previously produced output and applies the attention mechanism to it. This module applies *masked attention*. During the training, we will have pairs of input and output sequences that the model will learn from. It is important that the decoder learns to produce the next output by looking at only past tokens. It should not pay attention to the future tokens (otherwise, the whole point of developing a predictive model is moot). The masked attention module zeroes out the attention weights for the future tokens.

The second attention module takes in two inputs: the normalized output of the first module and the output from our encoder. The attention module has three inputs, known as the *query*, *key*, and *value* vectors. However, we will not go into specifics of these.

Finally, note that the decoder has an additional linear layer and a `softmax` layer at the end. The goal of the decoder is to produce sequences (mainly text) in the target language. Therefore, the embeddings that are generated must be somehow mapped to words. The `softmax` layer outputs a probability distribution over the vocabulary of tokens. The one with the maximum probability is chosen to be the output word.

Understanding BERT

So far, we have seen how the attention mechanism works and how the transformers leverage it for effective sequence-to-sequence modeling. As a final step, we will learn about BERT, a model that uses transformers and a novel set of training methodologies. The effectiveness of BERT and the utility of a pre-trained model in downstream tasks will be critical for our malware detection task.

BERT stands for **Bidirectional Encoder Representations from Transformers**. Ever since its introduction in 2018, it has made great impacts on the natural language processing world. It is a significant discovery that allows researchers and scientists to harness the power of large-scale machine learning language models, without the need for massive data or extensive computing resources.

BERT architecture

Architecturally speaking, BERT leverages transformers to create a structure. The BERT base model has 12 transformers stacked on top of each other and 12 self-attention heads. The BERT large model has 24 transformer layers and 16 self-attention heads. Both these models are *tremendously large* (with 110M and 40M parameters respectively).

Both of these models have been released by Google as open source and are freely available for anyone to use.

MLM as a training task

Traditional language models examine text sequences in only one direction: from left to right or right to left. This approach works just fine for generating sentences. The overarching goal in a language model is, given the words that have occurred so far, to predict the next word likely to appear.

However, BERT takes this a step further. Instead of looking at the sequence either left to right or right to left, it looks at the sequence both ways. It is trained on a task known as **masked language modeling** (**MLM**). The concept here is fairly simple. We randomly mask out some words (around 15%) from the sentence and replace them with a [MASK] token. Now, the goal is not to predict the next word in the sentence. The model will now learn to predict the masked words, given the surrounding words in both directions.

The word embeddings generated by traditional models represent words in a numeric space such that words similar in meaning are close to one another in the vector space. However, BERT will generate embeddings for a word depending on the context of the word. In such embeddings, the vector representation of a word changes with the context in which the word is being used.

In traditional word embeddings, a word will have the same embedding irrespective of the context. The word *match* will have the same embedding in the sentence "They were a perfect match" and "I lit a match last night." We clearly see that although the word is the same, the context matters and changes the meaning. BERT recognizes this and conditions the embeddings based on context. The word match has different embeddings in these two sentences.

Fine-tuning BERT

As mentioned before, the power of BERT lies in fine-tuning. The original BERT model has been trained on the masked language model task using the BooksCorpus data (containing 800 million words) and the Wikipedia data (containing 2.5 billion words). The model learns from large-scale datasets, training that we cannot reproduce trivially. For context, the BERT large model required 4 days and 16 cloud TPUs for training.

The concept of transfer learning helps us leverage this already-trained model for our downstream tasks. The idea behind this is that we take a generic language model, and fine-tune it for our specific task. High-level concepts are already learned by the model; we simply need to teach it more about a specific task.

In order to do this, we use the pre-trained model and add a single layer (often a single-layered neural network) on top of it. The nature of the layer will depend on the task we are fine-tuning for. For any task, we simply plug in the task-specific inputs and outputs in the correct format into BERT and fine-tune all the parameters end to end. As we fine-tune, we will achieve two tasks:

- The parameters of the transformer will be updated iteratively to refine the embeddings and generate task-specific contextual embeddings
- The newly added layer will be trained (that is, will learn appropriate parameters) to classify the new class of embeddings

Fine-tuning BERT is an inexpensive task both in terms of time and resources. Classification models can be built using an hour on a TPU, around 4 hours on a GPU, and 8-10 hours on a regular CPU. BERT has been used after fine-tuning several tasks such as question-answering, sentence completion, and text understanding.

Detecting malware with BERT

So far, we have seen attention, transformers, and BERT. But all of it has been very specific to language-related tasks. How is all of what we have learned relevant to our task of malware detection, which has nothing to do with language? In this section, we will first discuss how we can leverage BERT for malware detection and then demonstrate an implementation of the same.

Malware as language

We saw that BERT shows excellent performance on sentence-related tasks. A sentence is merely a sequence of words. Note that we as humans find meaning in a sequence because we understand language. Instead of words, the tokens could be anything: integers, symbols, or images. So BERT performs well on sequence tasks.

Now, imagine that instead of words, our tokens were calls made by an application. The life cycle of an application could be described as a series of API calls it makes. For instance, `<START> <REQUEST-URL> <DOWNLOAD-FILE> <EXECUTE-FILE> <OPEN-CONTACTS> <POST-URL> <END>` could represent the behavior of an application. Just like a sentence is a sequence of words, an application can be thought of as a sequence of API calls.

The relevance of BERT

Recall that BERT learned contextual embeddings for a word. If we use BERT with malware data (with API calls as words and their sequence as sentences), the model will be able to learn embedding representations for each API call and condition it on the context. This is useful for malware detection because a single API call cannot determine whether an application is malware or not, but the context in which it is called might. For example, by itself, the API call for making a request to a third-party

URL may not be malicious, but coupled with accessing stored passwords and contacts, it may indicate malware at work. This is our motivation behind choosing BERT as a model for malware detection.

We will use BERT on the malware classification task just like a sentence classification task. Here we have two options:

- Train a new BERT model from scratch
- Fine-tune the existing BERT model (pre-trained on language data) on malware classification

As the domains of the pre-trained model are different from malware (that is, tokens in malware API sequences will not appear in the Wikipedia or BooksCorpus datasets), the first option might seem better. However, recall that the datasets for pre-training were massive, and we do not have access to malware data at that scale.

Prior research has shown that a BERT model, even when pre-trained in the English language, serves as an excellent candidate for malware detection. This is because the pre-training results in an optimal set of parameters that results in faster convergence of any downstream task such as malware detection. In the following sections, this is the approach we will take. We will first preprocess the data, read it into a DataFrame, and then fine-tune it on a pre-trained BERT model.

Getting the data

As described previously, malware can be analyzed using both static and dynamic methods. Several commercial tools exist for decompiling malware binaries, understanding the behavior, and examining their activities. One such tool for malware analysis is WildFire (`https://www.paloaltonetworks.com/products/secure-the-network/wildfire`), developed by Palo Alto Networks. It is a cloud malware protection engine that utilizes advanced machine learning models to detect targeted malware attacks in real time.

Creating your own datasets for malware detection is challenging. First, using tools such as WildFire to generate dynamic analysis files is an expensive task (commercial tools are generally patented and require a license) and also outside the scope of this book. Second, examples of malware, particularly those seen in the wild, are hard to come by. Finally, experimenting with malware executables may inadvertently infect your systems. We will, therefore, use a commercially available malware dataset.

In 2018, Palo Alto Networks released a research paper (`https://arxiv.org/pdf/1812.07858.pdf`) that discussed common cybersecurity problems, in which malware detection was discussed in great detail. Along with the paper, they released a malware dataset that contained analysis files for over 180,000 different applications. The dataset is made of a sample of malware identified by Palo Alto Networks in a given period. For each malware, they provide identifiers and the domains accessed by the file, along with the sequence of API calls made by the application.

The dataset is freely available for students and researchers and can be obtained by contacting Palo Alto Networks as described in the paper. However, the method we present is fairly generic and can be applied to any malware dataset that you can gain access to.

Preprocessing the data

The Palo Alto Networks dataset contains several features for every application, including static and dynamic analysis files. However, of particular interest to us is the API call sequence. This is because we want to exploit the power of transformers. A defining characteristic of transformers is that they excel at handling sequential data via attention mechanisms.

The dynamic analysis file will give us the sequence of API calls made by the application. Each API call consists of two parts: the **action** and the **key**. The action refers to the actual task behind the API call (such as opening a file, turning on Bluetooth, or sending a GET request). The key refers to the parameters passed to the API call (such as the actual domain to which the application is connected). While examining the parameters will also reveal significant information about whether an app is malicious, here we simply focus on the action of the API call.

Every application (whether malware or benign) has an associated XML file that contains the API call logs. Once they are extracted, we will have access to the set of actions taken by the application. An example snippet could look like this:

```
APK file used SSL
APK file set up an alarm
APK file got a system property
APK file read a file on SDC
APK file connected to host
APK file set up an alarm
APK file requested BT access
...
```

Figure 3.5 – API call sequence

We first extract the sequence for all the applications. Now that we have the sequence, it must be converted into tokens suitable for consumption by a machine learning model. Once we have all the sequences, we can compute the universe of known API calls, and assign a unique integer to each API call. Every application can now be represented as a sequence of integers.

Note that the steps discussed so far are not specific to the Palo Alto Networks dataset, and this is the reason why we did not introduce any specific functions or code to do the preprocessing. You can apply this same technique to construct a feature vector from any malware dataset.

For convenience and simplicity, we will provide the preprocessed versions of the dataset. Note that the API call sequences are represented by integers, as we cannot share the exact API calls publicly. Interested readers can refer to the original paper and data if they are curious to learn more.

Building a classifier

We will leverage the BERT model to build our classifier. Much of the code here is borrowed from the official notebooks released by Google when they released the BERT model in 2019. Some of the code may seem intimidating; however, do not worry. A lot of this is just boilerplate environment setup and function definitions that you absolutely do not need to understand in detail. We will go over the code and discuss what parts need to be changed when you implement this on your own or wish to use this setup for a different problem.

First, we will import the required libraries. If any of these are not already installed (and Python will throw an error saying so), then they can be installed using the `pip` utility:

```
import bert
from bert import run_classifier
from bert import optimization
from bert import tokenization
import tensorflow as tf
import numpy as np
import pandas as pd
import tensorflow_hub as hub
```

We will leverage a pre-trained BERT model for building our classifier. TensorFlow Hub contains all of these pre-trained models and they are available for use by the public. This function reads the model and its vocabulary and generates a tokenizer. The tokenizer is responsible for converting the words we see into tokens that can be understood by the machine learning model:

```
BERT_MODEL_HUB = "https://tfhub.dev/google/bert_cased_L-12_H-
768_A-12/1"

def create_tokenizer_from_hub_module():
  with tf.Graph().as_default():
    bert_module = hub.Module(BERT_MODEL_HUB)
    tokenization_info = bert_module(signature="tokenization_info",
as_dict=True)
  with tf.Session() as sess:
    vocab_file, do_lower_case = sess.run([tokenization_info["vocab_
file"],
                                          tokenization_
info["do_lower_case"]])

  return bert.tokenization.FullTokenizer(
      vocab_file=vocab_file, do_lower_case=do_lower_case)

tokenizer = create_tokenizer_from_hub_module()
```

Now, we actually create the classification model. This function will create a BERT module and define the input and output structures. We will then obtain the output layer and find the parameters there; these are the ones that would be used to run inference. We apply a dropout to this layer and calculate the logits (that is, the `softmax` output of the layer). During the training phase, the `softmax` output is used to compute the loss relative to the ground truth. In the inferencing phase, the output can be used to predict the token depending on which probability in the `softmax` output is the highest. Note that the ground truth is in the form of categorical tokens (in our case, API calls) and, therefore, needs to be converted into a numeric form using one-hot encoding:

```python
def create_model(is_predicting, input_ids, input_mask, segment_ids,
labels,num_labels):
  """Creates a classification model."""

  bert_module = hub.Module(
      BERT_MODEL_HUB,
      trainable=True)
  bert_inputs = dict(
      input_ids=input_ids,
      input_mask=input_mask,
      segment_ids=segment_ids)
  bert_outputs = bert_module(
      inputs=bert_inputs,
      signature="tokens",as_dict=True)
  output_layer = bert_outputs["pooled_output"]
  hidden_size = output_layer.shape[-1].value
  output_weights = tf.get_variable("output_weights", [num_
labels, hidden_size],      initializer=tf.truncated_normal_
initializer(stddev=0.02))

  output_bias = tf.get_variable("output_bias",[num_labels],
initializer=tf.zeros_initializer())

  with tf.variable_scope("loss"):

    # Dropout helps prevent overfitting
  output_layer = tf.nn.dropout(output_layer, keep_prob=0.9)
    logits = tf.matmul(output_layer, output_weights, transpose_b=True)
    logits = tf.nn.bias_add(logits, output_bias)
    log_probs = tf.nn.log_softmax(logits, axis=-1)

    # Convert labels into one-hot encoding
    one_hot_labels = tf.one_hot(labels, depth=num_labels, dtype=tf.
float32)
```

```
    predicted_labels = tf.squeeze(tf.argmax(log_probs, axis=-1,
output_type=tf.int32))
    # If we're predicting, we want predicted labels and the
probabilties.
    if is_predicting:
      return (predicted_labels, log_probs)

    # If we're train/eval, compute loss between predicted and actual
label
    per_example_loss = -tf.reduce_sum(one_hot_labels * log_probs,
axis=-1)
    loss = tf.reduce_mean(per_example_loss)
    return (loss, predicted_labels, log_probs)
```

The code so far has been mostly boilerplate functions for setting up the parameters. Now, we will create a function that actually defines our training and inference settings. Recall that the previous function defined the steps to create the model. This function will leverage the previous one for training and inference.

First, we read the input features and labels that are crucial to training. These will be passed to the function as parameters. If the training phase is going on, the function will use the `create_model` function to calculate a loss that will be optimized for training. If not, it will simply score the data point on the model and return a predicted label and output probabilities.

This function also has a metric calculation function defined. This is crucial to analyzing and comparing the performance of our model. TensorFlow has built-in functions that calculate common metrics such as precision, recall, false positives, false negatives, F1 score, and so on. We leverage these built-in functions and return a dictionary, which contains various metrics:

```
def model_fn_builder(num_labels, learning_rate, num_train_steps,num_
warmup_steps):
  def model_fn(features, labels, mode, params):

    input_ids = features["input_ids"]
    input_mask = features["input_mask"]
    segment_ids = features["segment_ids"]
    label_ids = features["label_ids"]

    is_predicting = (mode == tf.estimator.ModeKeys.PREDICT)

    # TRAIN and EVAL
    if not is_predicting:
      (loss, predicted_labels, log_probs) = create_model(is_
predicting, input_ids, input_mask, segment_ids, label_ids, num_labels)
```

```
      train_op = bert.optimization.create_optimizer(loss, learning_
rate, num_train_steps, num_warmup_steps, use_tpu=False)

   # Calculate evaluation metrics.
   def metric_fn(label_ids, predicted_labels):
     accuracy = tf.metrics.accuracy(label_ids, predicted_labels)
     f1_score = tf.contrib.metrics.f1_score(
         label_ids,
         predicted_labels)
     auc = tf.metrics.auc(
         label_ids,
         predicted_labels)
     recall = tf.metrics.recall(
         label_ids,
         predicted_labels)
     precision = tf.metrics.precision(
         label_ids,
         predicted_labels)
     true_pos = tf.metrics.true_positives(
         label_ids,
         predicted_labels)
     true_neg = tf.metrics.true_negatives(
         label_ids,
         predicted_labels)
     false_pos = tf.metrics.false_positives(
         label_ids,
         predicted_labels)
     false_neg = tf.metrics.false_negatives(
         label_ids,
         predicted_labels)
     return {
         "eval_accuracy": accuracy,
         "f1_score": f1_score,
         "auc": auc,
         "precision": precision,
         "recall": recall,
         "true_positives": true_pos,
         "true_negatives": true_neg,
         "false_positives": false_pos,
         "false_negatives": false_neg
     }

   eval_metrics = metric_fn(label_ids, predicted_labels)
```

```
        if mode == tf.estimator.ModeKeys.TRAIN:
          return tf.estimator.EstimatorSpec(mode=mode,
            loss=loss,
            train_op=train_op)
        else:
          return tf.estimator.EstimatorSpec(mode=mode,
              loss=loss,
              eval_metric_ops=eval_metrics)
        else:
          (predicted_labels, log_probs) = create_model(
          is_predicting, input_ids, input_mask, segment_ids, label_ids,
   num_labels)

          predictions = {
              'probabilities': log_probs,
              'labels': predicted_labels
          }
        return tf.estimator.EstimatorSpec(mode, predictions=predictions)

    # Return the actual model function in the closure
    return model_fn
```

Now that all of the functions to initialize the model, compute loss and training, and inferencing are good to go, we can use this setup by plugging in our malware data. Note that the procedure so far is generically applicable to any fine-tuning problem you are trying to solve using BERT; nothing has been hardcoded or is specific to malware data. If you want to use the fine-tuning approach to BERT on another task (sentiment analysis, hate speech detection, or misinformation detection), all of the steps we have completed so far remain valid.

We will now define some parameters used for training. The first set of parameters are standard machine learning ones. The batch size defines the number of examples that will be used to calculate loss at a time, and the learning rate defines the rate at which parameters will be updated in the gradient descent optimization algorithm. The number of epochs is set here to 3, which is a small number. This is because we are not training a model from scratch; we are simply using an already trained model, and fine-tuning it to operate on our dataset.

The next set of parameters exists for optimization and ease in training. It defines after how many steps a new version of the model should be saved. Here, we have set it to 500, meaning that after every 500 steps, a new model will be saved. This helps us if we run into an unexpected error or crash; the model simply reads the latest saved model and picks up training from that point.

Finally, the last set of parameters defines the positive ratio. This is the proportion of malware samples in the training data. Here we set it to 0.001, which amounts to 0.1%:

```
BATCH_SIZE = 32
LEARNING_RATE = 2e-5
NUM_TRAIN_EPOCHS = 3.0
WARMUP_PROPORTION = 0.1

SAVE_CHECKPOINTS_STEPS = 500
SAVE_SUMMARY_STEPS = 100

POSRATIO=0.001 # 0.1%
NPOS=10000*POSRATIO
```

Now, we read the data frame that contains our data. Recall that the VERDICT column contained a 0/1 label indicating whether that particular data point was malware or not. We separate the malware and benign samples, sample the required fraction from the positive class, and then combine it with the negative class. This way, we have a dataset that contains only the required proportion of malware samples:

```
df_a=pd.read_csv('dataset.csv')
df_a_1=df_a[df_a['VERDICT']==1]
df_a_0=df_a[df_a['VERDICT']==0]
df_a_sampled=pd.concat([df_a_1[:nPos],df_a_0[:NPOS]])
```

We now split our data into training and testing sets. Note that here we have a very small proportion of malware in our sampled data. If we split randomly, we might end up with all of the malware entirely in the training or testing set. To avoid this, we apply stratified sampling. With stratified sampling, the proportion of labels remains roughly the same in both the training and testing datasets:

```
from sklearn.model_selection import train_test_split
df_train_n,df_test_n=train_test_split(df_a_sampled,stratify=df_a_
sampled['VERDICT'])
```

Remember that the data we have is in the form of API call sequences. This has to be converted into a form suitable for being consumed by BERT. We do this in our next step and transform both the training and test data into the required format.

First, we use the InputExample class to wrap our API sequence string and labels together:

```
DATA_COLUMN='API_CALL_SEQ'
LABEL_COLUMN='VERDICT'
train InputExamples_a = df_train_n.apply(lambda x: bert.run_
classifier.InputExample(guid None,
                        text_a = x[DATA_COLUMN],
                        text_b = None,
                        label = x[LABEL_COLUMN]),
```

```
                                        axis = 1)

test_InputExamples_a = df_test_n.apply(lambda x: bert.run_classifier.
InputExample(guid=None,
                        text_a = x[DATA_COLUMN],
                        text_b = None,
                        label = x[LABEL_COLUMN]),
                        axis = 1)
```

Then, we transform the sequence into features using our tokenizer:

```
label_list=[0,1]
MAX_SEQ_LENGTH = 128

train_features_a = bert.run_classifier.convert_examples_to_features(
                train_InputExamples_a,
                label_list,
                MAX_SEQ_LENGTH,
                tokenizer)
test_features_a = bert.run_classifier.convert_examples_to_features(
                test_InputExamples_a,
                label_list,
                MAX_SEQ_LENGTH,
                tokenizer)
```

We now have our features and labels. We are ready to train the model! We will use the model function builder we defined earlier to create the model and pass it to the TensorFlow estimator, which will take care of the training for us. We specify the output directory in which to save the trained model as well as the parameters we defined earlier (summary and checkpoint steps) in a run configuration. This also gets passed to the estimator:

```
OUTPUT_DIR='saved_models/rate_'+str(posRatio*100)
num_train_steps = int(len(train_features_a) / BATCH_SIZE * NUM_TRAIN_
EPOCHS)
num_warmup_steps = int(num_train_steps * WARMUP_PROPORTION)

run_config = tf.estimator.RunConfig(
    model_dir=OUTPUT_DIR,
    save_summary_steps=SAVE_SUMMARY_STEPS,
    save_checkpoints_steps=SAVE_CHECKPOINTS_STEPS)

model_fn = model_fn_builder(
  num_labels=len(label_list),
  learning_rate=LEARNING_RATE,
```

```
  num_train_steps=num_train_steps,
  num_warmup_steps=num_warmup_steps)

estimator_android = tf.estimator.Estimator(
  model_fn=model_fn,
  config=run_config,
  params={"batch_size": BATCH_SIZE})

train_input_fn_a = bert.run_classifier.input_fn_builder(
    features=train_features_a,
    seq_length=MAX_SEQ_LENGTH,
    is_training=True,
    drop_remainder=False)

import time
print(f'Beginning Training!')
current_time = time.time()
estimator_android.train(input_fn=train_input_fn_a, max_steps=num_
train_steps)
print("Training took time ", time.time() - current_time)
```

This will produce a lot of output, most of which you do not need to understand. The training time will vary depending on the processor you are using, the GPU (if any), and system usage. Without a GPU, the fine-tuning took approximately 29,000 seconds, which amounts to roughly eight hours.

Finally, we want to use this fine-tuned model to make predictions for new data and evaluate its performance. We can use the same estimator in inference mode:

```
test_input_fn_a = bert.run_classifier.input_fn_builder(
    features=test_features_a,
    seq_length=MAX_SEQ_LENGTH,
    is_training=False,
    drop_remainder=False)

metrics = estimator_android.evaluate(input_fn=test_input_fn_a,
steps=None)
```

This should show you an output that prints out the metrics. Note that your numbers may differ slightly from the ones you see here:

```
{'auc': 0.95666675,
 'eval_accuracy': 0.99920031,
 'f1_score': 0.49999997,
 'false_negatives': 2.0,
 'false_positives': 0.0,
```

```
'loss': 0.0076462436,
'precision': 0.974,
'recall': 0.871,
'true_negatives': 2500.0,
'true_positives': 1.0,
'global_step': 703}
```

Recall that earlier, we defined a model evaluation function. The variable metrics will contain the dictionary with the various evaluation metrics. If you print it out, you should be able to examine the accuracy, precision, recall, and F-1 score.

This completes our experiment! We have successfully used a BERT model pre-trained on a language task and fine-tuned it to classify malicious applications based on the API call sequence. Feel free to experiment with the code. Here are some things to ponder:

- What happens if you use the BERT large model instead of the BERT base model?
- How does the performance vary with the positive rate fraction?
- What happens if you vary the architecture (add more layers)?

With that, we have come to the end of this chapter.

Summary

This chapter provided an introduction to malware and a hands-on blueprint for how it can be detected using transformers. First, we discussed the concepts of malware and the various forms they come in (rootkits, viruses, and worms). We then discussed the attention mechanism and transformer architecture, which are recent advances that have taken the machine learning world by storm. We also looked at BERT, a model that has beat several baselines in tasks such as sentence classification and question-answering. We leveraged BERT for malware detection by fine-tuning a pre-trained model on API call sequence data.

Malware is a pressing problem that places users of phones and computers at great risk. Data scientists and machine learning practitioners who are interested in the security space need to have a strong understanding of how malware works and the architecture of models that can be used for detection. This chapter provided all of the knowledge needed and is a must to master for a SecML professional.

In the next chapter, we will switch gears and turn to a different problem: fake online reviews.

4

Detecting Fake Reviews

Reviews are an important element in online marketplaces as they convey the customer experience and their opinions on products. Customers heavily depend upon reviews to determine the quality of a product, the truth about various claims in the description, and the experiences of other fellow customers. However, in recent times, the number of fake reviews has increased. Fake reviews are misleading and fraudulent and cause harm to consumers. They are prevalent not only on shopping sites but also on any site where there is a notion of reputation through reviews, such as Google Maps, Yelp, Tripadvisor, and even the Google Play Store.

Fraudulent reviews harm the integrity of the platform and allow scammers to profit, while genuine users (sellers and customers) are harmed. As data scientists in the security space, understanding reputation manipulation and how it presents itself, as well as techniques for detecting it, is essential. This chapter focuses on examining reputation manipulation through fake reviews.

In this chapter, we will cover the following main topics:

- Reviews and integrity
- Statistical analysis
- Modeling fake reviews with regression

By the end of this chapter, you will have a clear understanding of reputation manipulation through fake reviews and how they can be detected. You will also learn about statistical tests and how to apply them for analysis and how the reviews data can be modeled using regression.

Technical requirements

You can find the code files for this chapter on GitHub at https://github.com/ PacktPublishing/10-Machine-Learning-Blueprints-You-Should-Know-for-Cybersecurity/tree/main/Chapter%204

Reviews and integrity

Let us first look at the importance of online reviews and why fake reviews exist.

Why fake reviews exist

E-commerce websites always have reviews for products. Reviews play an important role in the online world. Reviews allow consumers to post their experiences and facilitate peer-to-peer reputation building. Reviews are important on online platforms for several reasons:

- Online reviews provide valuable information to potential customers about the quality and performance of a product or service. Customers can read about other people's experiences with a product or service before deciding whether to buy it or not.

- Reviews from other customers help build trust between the seller and the buyer. Positive reviews can reassure potential customers that a product or service is worth buying, while negative reviews can warn them about potential problems.

- Online reviews can provide businesses with valuable feedback about their products and services. This feedback can help businesses improve their offerings and customer service.

- Online reviews can also help businesses improve their search engine rankings. Search engines such as Google take into account the number and quality of reviews when ranking websites in search results.

It is therefore natural that better reviews imply better sales and more profit for the seller. The seller has an incentive to have as many great reviews for their product as possible, and this has led to the problem of fake reviews.

Fake reviews are reviews that are deliberately written to mislead or deceive readers. They can be positive or negative and are usually written by individuals or companies who have a vested interest in manipulating the reputation of a product or service. Here are some common types of fake reviews:

- **Paid reviews**: Some individuals or companies pay people to write positive reviews about their products or services, even if they haven't used them

- **Fake negative reviews**: Competitors or individuals with a grudge may write fake negative reviews to harm the reputation of a business or product

- **Review swaps**: Some individuals or companies offer to exchange positive reviews with other businesses or individuals to boost their own ratings

- **Review bots**: Some businesses use automated software programs to generate large numbers of fake reviews quickly

Fake reviews are problematic because they can mislead potential customers into making poor purchasing decisions. They can also harm the reputation of businesses that rely on genuine customer feedback to improve their products and services. Many online platforms have policies in place to detect and remove fake reviews to protect the integrity of their review systems. For example, Amazon explicitly bans reviewers from receiving any compensation or free products in exchange for reviews.

Evolution of fake reviews

The problem of fake reviews and reputation manipulation is not a new one – it has existed for decades. However, the nature of fake reviews has changed significantly, from bot-generated reviews to crowdsourced reviews and then to incentivized reviews.

Bot-generated reviews

These were the very first form of fake reviews that could be onboarded by sellers at scale. Bot-generated reviews are reviews that are created using automated software programs, also known as review bots. These bots are designed to generate and post a large number of reviews quickly, without any human intervention. These reviews have several tell-tale signs:

- They are usually generic and lack specific details about the product or service being reviewed
- They may use similar language and sentence structures and often have a high number of positive ratings
- They also exhibit similarities in terms of the IP addresses, subnets, and networks that they come from

Many online platforms have implemented measures to detect and remove bot-generated reviews to maintain the integrity of their review systems. Some of these measures include using machine learning algorithms to identify patterns in the reviews, monitoring for suspicious IP addresses and activity, and requiring reviewers to verify their identities.

Crowdsourced reviews

Knowing that bot-generated reviews were easy to detect by flagging IP addresses and other symptoms of automated activity, malicious sellers turned to crowdsourced reviews. With this kind of fake reviews, crowd workers are hired simply to write hundreds of reviews for products. These individuals are often part of online marketplaces or platforms that offer payment in exchange for writing positive reviews about products or services. Crowd workers may not have actually used the product or service being reviewed and may simply be provided with basic information to write a review. Sourced on freelancing and crowd-working websites such as Amazon MTurk or Fiverr, these crowd workers work on a commission basis and earn a fee for every review they post. Notably, these reviews have certain peculiar characteristics as well. As many reviews are written by the same user, they show very high inter-review similarity. They also show high burstiness (that is, a sudden spike in the number of reviews, corresponding to the time when the crowd worker was hired). These signals can be used as features to detect this type of reviews.

Incentivized reviews

These are the latest trend in online fake reviews. Incentivized reviews are reviews written by customers who have received some form of compensation or reward in exchange for their reviews. This compensation can come in various forms, such as free products, discounts, gift cards, refunds, or other incentives. Incentivized reviews are often used by companies as a marketing strategy to increase positive reviews and ratings of their products or services. By providing incentives to customers, companies hope to encourage them to write positive reviews, which can then be used to attract more customers and boost sales.

However, incentivized reviews can be controversial because they can create a bias toward positive reviews and may not accurately reflect the true opinions of customers. This is because customers who receive incentives may feel obligated to write a positive review, even if they did not have a positive experience with the product or service. As a result, many review platforms and websites have strict policies against incentivized reviews and may remove them from their platforms to ensure the integrity of their review system.

Compared to bot-generated or crowdsourced reviews, incentivized reviews are less easy to detect. These are written by human users and hence do not show the symptoms of automated activity as bots do. As these are real users and not crowd workers, these reviews also show less similarity.

Statistical analysis

In this section, we will try to understand some review data and check whether there are any differences between genuine and fake reviews. We will use the Amazon fake reviews dataset that Amazon has published on Kaggle. It is a set of around 20,000 reviews with associated labels (real or fake) as labeled by domain experts at Amazon.

Exploratory data analysis

We will first load up the data and take a first pass over it to understand the features and their distribution.

We begin by importing the necessary libraries:

```
import numpy as np
import pandas as pd
import matplotlib.pyplot as plt
```

We will then read the `reviews` data. Although it is a text file, it is structured and therefore can be read with the `read_csv` function in Pandas:

```
reviews_df = pd.read_csv("amazon_reviews.txt", sep="\t")
reviews_df.head()
```

This is what the output should look like:

DOC_ID		LABEL	RATING	VERIFIED_PURCHASE	PRODUCT_CATEGORY	PRODUCT_ID	PRODUCT_TITLE	REVIEW_TITLE	REVIEW_TEXT
0	1	__label1__	4	N	PC	B00008NG7N	Targus PAUK10U Ultra Mini USB Keypad, Black	useful	When least you think so, this product will sav...
1	2	__label1__	4	Y	Wireless	B00LH0Y3NM	Note 3 Battery : Stalion Strength Replacement ...	New era for batteries	Lithium batteries are something new introduced...
2	3	__label1__	3	N	Baby	B000I5UZ1Q	Fisher-Price Papasan Cradle Swing, Starlight	doesn't swing very well.	I purchased this swing for my baby. She is 6 m...
3	4	__label1__	4	N	Office Products	B003822IRA	Casio MS-80B Standard Function Desktop Calculator	Great computing!	I was looking for an inexpensive desk calculat...
4	5	__label1__	4	N	Beauty	B00PWSAXAM	Shine Whitening - Zero Peroxide Teeth Whitenin...	Only use twice a week	I only use it twice a week and the results are...

Figure 4.1 – A glimpse of the reviews dataset

Notice the **LABEL** column in the data. Rather than a simple label, it has labels of **__label1__** and **__label2__**. Looking at the documentation for this dataset, we can see that **__label1__** corresponds to real reviews, and **__label2__** to fake ones.

For easier understanding, we will transform these labels. We want 0 to correspond to a real review, and 1 to a fake one. The following code snippet does this for us:

```
def label_to_int(label):
    if label == "__label2__":
        # Real Review
        return 0
    else:
        # Fake Review
        return 1

reviews_df["FRAUD_LABEL"] = reviews_df["LABEL"].apply(label_to_int)
reviews_df.head()
```

The output of this code is as follows. You can see that a new column, **FRAUD_LABEL**, with **0** and **1** values has been created.

DOC_ID		LABEL	RATING	VERIFIED_PURCHASE	PRODUCT_CATEGORY	PRODUCT_ID	PRODUCT_TITLE	REVIEW_TITLE	REVIEW_TEXT	FRAUD_LABEL
0	1	__label1__	4	N	PC	B00008NG7N	Targus PAUK10U Ultra Mini USB Keypad, Black	useful	When least you think so, this product will sav...	1
1	2	__label1__	4	Y	Wireless	B00LH0Y3NM	Note 3 Battery : Stalion Strength Replacement ...	New era for batteries	Lithium batteries are something new introduced...	1
2	3	__label1__	3	N	Baby	B000I5UZ1Q	Fisher-Price Papasan Cradle Swing, Starlight	doesn't swing very well.	I purchased this swing for my baby. She is 6 m...	1
3	4	__label1__	4	N	Office Products	B003822IRA	Casio MS-80B Standard Function Desktop Calculator	Great computing!	I was looking for an inexpensive desk calculat...	1
4	5	__label1__	4	N	Beauty	B00PWSAXAM	Shine Whitening - Zero Peroxide Teeth Whitenin...	Only use twice a week	I only use it twice a week and the results are...	1

Figure 4.2 – Dataset with clear labels added

First, we want to look at the distribution of real and fake reviews. This will tell us whether the dataset is balanced. If it is an imbalanced dataset, we may have problems in building a classifier. If it is highly imbalanced (such as only 1% of the reviews being fake), we may want to move from classification to an anomaly detection approach.

Note that there is a product category feature in the dataset. We want to examine how reviews are distributed across categories. We do this because different kinds of products may have different kinds of reviews. The nature of fake reviews for home apparel might be different from the ones for electronics products. We want to look at the distribution to anticipate any bias or generalization concerns.

To do so, we will group the reviews by category, and count the number of reviews per category. The result in the form of a bar graph will show us the distribution:

```
axes = reviews_df.groupby("FRAUD_LABEL").PRODUCT_CATEGORY\
       .value_counts()\
       .unstack(0)\
       .plot.barh()
axes.set_xlabel("# Reviews")
axes.set_ylabel("Product Category")
```

The output should be as follows. We can see that there are ~30 categories of products, and each one has 350 real and 350 fake reviews. This dataset is therefore well balanced and we do not need to worry about bias coming from the review category feature.

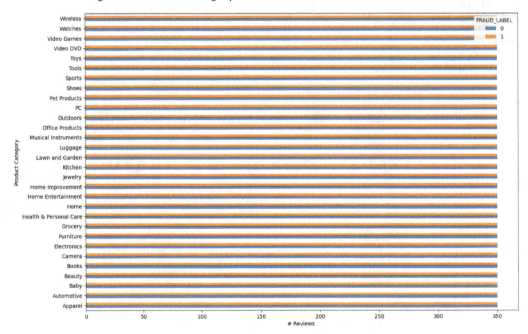

Figure 4.3 – Review distribution across categories

Next, we want to check how the real and fake reviews are distributed across ratings. This is an important factor to consider, and failing to do so can result in a biased model. For example, if the dataset is set up so that the fake reviews are all 4-star and 5-star and all the real reviews are 1-star and 2-star, our model will learn to detect sentiment instead of review authenticity. The sentiment will introduce bias into our model and this beats the actual purpose of building the model itself.

We will group the reviews by rating and calculate the number of real and fake reviews for every rating. Note that this is feasible because there are only five classes of ratings and they are discrete:

```
axes = reviews_df.groupby("FRAUD_LABEL").RATING\
        .value_counts()\
        .unstack(0)\
        .plot.bar()
axes.set_xlabel("# Reviews")
axes.set_ylabel("Rating")
```

The output is shown as follows. We can see that there is some disparity in the number of reviews by rating class (there are only 2,000 1-star reviews whereas there are around 12,000 5-star reviews). However, within each class, the number of real and fake reviews is roughly equal. Therefore, the data is well distributed on the rating front as well.

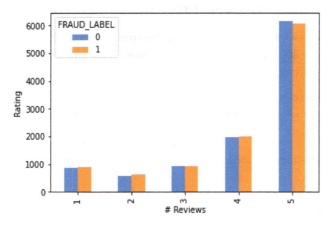

Figure 4.4 – Review distribution across ratings

Finally, we will review the distribution of real and fake reviews by whether they are verified or not. A review being verified means that the e-commerce platform guarantees that the product was actually purchased by the reviewer, typically on the platform itself. We can observe the distribution with the following code snippet:

```
axes = reviews_df.groupby("FRAUD_LABEL").VERIFIED_PURCHASE\
        .value_counts()\
        .unstack(0)\
        .plot.bar()
axes.set_xlabel("Purchase Verified")
axes.set_ylabel("Rating")
```

The output is shown in the following bar plot:

Figure 4.5 – Review distribution across the verified label

Looking at it, we observe two interesting occurrences:

- In the verified reviews, the percentage of genuine (real) reviews is higher than the fake ones. There are almost 9,000 real reviews, but only around 3,000 fake ones. This means that only 25% of verified reviews are fake.

- In the non-verified reviews, we see that the trend is reversed. There are almost 7,000 fake reviews and approximately 1,800 real reviews. Therefore, almost 80% of non-verified reviews are fake.

This trend is natural. It is simpler for a bot or review service to trivially generate non-verified reviews. However, generating a verified review involves actually buying the product, which incurs additional expenses.

Feature extraction

In this section, we will extract some features from the `reviews` data. The goal is to characterize fake reviews by certain signals or trends they may exhibit. Here, we build these features using our intuition, as well as leveraging prior research in the field.

First, we will write a function to extract the length of the review. This function first checks whether the review text is empty. If so, it returns 0. If not, it returns the number of words in the review:

```
def review_length(text):
    if text is None:
        return 0
    else:
        words = text.split(" ")
        return len(words)
```

Next, we will write a function to compute the average word length in the review. To do so, we will split the review into individual words, add up their lengths, and divide by the total number of words to compute the mean. Of course, we must first check that the review is not empty:

```
def average_word_length(text):
    if text is None or text == "":
        return 0
    else:
        words = text.split(" ")
        total_lengths = 0
        for word in words:
            total_lengths = total_lengths + len(word)

        avg_len = total_lengths/len(words)
        return avg_len
```

Another feature we will derive is the number of words spelled incorrectly in the review. We will use the enchant Python library for this. The enchant Python library is a module that provides an interface to the enchant spellchecking system. enchant is a C/C++ library that provides a unified interface for several different spellchecking engines, including Aspell, Hunspell, and MySpell. enchant can be used from within your Python code to perform spellchecking on text. enchant provides a number of useful features for spellchecking, including suggestions for misspelled words, the ability to add and remove words from a user dictionary, and the ability to work with multiple languages.

To use enchant in your Python code, you first need to install the enchant library on your system. This can be done using the Python pip utility. Once installed, you can import the enchant module into your Python code and use the provided functions to perform spell checking.

To derive our feature, we will split the review into words, and count the number of words that are incorrectly spelled:

```
import enchant
def count_misspellings(text):
    english_dict = enchant.Dict("en_US")
    if text is None or text == "":
        return 0
    else:
        misspelling = 0
        words = text.split(" ")
        for word in words:
            if word != "" and not english_dict.check(word.lower()):
                misspelling = misspelling + 1
        return misspelling
```

Finally, now that we have defined the functions to extract features, it is time to put them to use. We will efficiently compute the features for each review in our DataFrame using the `apply()` function in Pandas.

The `apply()` function in the pandas library is used to apply a given function to every element of a pandas DataFrame, Series, or column. It can be used to transform, filter, or aggregate the data in the DataFrame, and is a powerful tool for data manipulation.

The `apply()` function takes a function as an argument, which is applied to every element in the DataFrame. The function can be defined by the user or can be a built-in Python function. In this case, we will use the custom functions we defined earlier. The `apply()` function also has some optional arguments that can be used to customize its behavior, such as its axis (to specify whether to apply the function row-wise or column-wise) and arguments (to pass additional arguments to the function).

Here is how we will use it:

```
reviews_df["Review_Text_Length"] = reviews_df["REVIEW_TEXT"].
apply(review_length)
reviews_df["Avg_Word_Len"] = reviews_df["REVIEW_TEXT"].apply(average_
word_length)
reviews_df["Num_Misspelling"] = reviews_df["REVIEW_TEXT"].apply(count_
misspellings)
```

This completes our feature-engineering code. You can, of course, add more features if you wish to experiment!

Statistical tests

Now that we have our features, we want to check whether they differ between real and fake reviews. The process of testing for differences is known as hypothesis testing.

Hypothesis testing

Hypothesis testing is a statistical method used to determine whether a claim or hypothesis about a population parameter is supported by the evidence provided by a sample of data. In other words, it is a way to test the validity of a hypothesis or claim about a population based on a sample of data.

The hypothesis being tested is typically called the null hypothesis (H0), and the alternative hypothesis (H1) is the hypothesis that is considered as an alternative to the null hypothesis. The null hypothesis usually represents the status quo or the default assumption, and the alternative hypothesis represents the hypothesis that the researcher is trying to support.

From the previous section, we have data belonging to two different groups: real reviews and fake reviews. Here is how the process of hypothesis testing will work for us:

1. **Formulating the null and alternative hypotheses**: The null hypothesis represents the status quo or the default assumption, and the alternative hypothesis represents the hypothesis that the researcher is trying to support. In this case, the null hypothesis would be that there is no difference between the features of the group of fake reviews and the group of real reviews. The alternative hypothesis would be that there is a difference between the features of the two groups.

2. **Collecting and analyzing data**: A sample of data is collected, and descriptive statistics and inferential statistics are used to analyze the data. A sample of reviews is collected, including both fake and real reviews. The feature of interest (such as the length or average word length) is computed for each review in the sample, and descriptive statistics are calculated for each group of reviews, including the mean and standard deviation of the feature.

3. **Choosing the level of significance**: The level of significance represents the probability that we will reject the null hypothesis even when it is true. The lower the significance level, the lower the chances of falsely rejecting the null hypothesis, which means higher confidence in our estimate. The most commonly used level of significance is 0.05, which means that there is a 5% chance of rejecting the null hypothesis when it is actually true.

4. **Calculating the test statistic**: A test statistic is calculated based on the sample data, which measures how far the sample statistic is from the null hypothesis. The test statistic depends on the type of test being performed. For example, if we are comparing the means of the two groups, we can use a t-test to calculate the t-value, which measures the difference between the means of the fake reviews and the real reviews relative to the variability of the data.

5. **Determining the p-value**: The p-value is the probability of obtaining a test statistic as extreme or more extreme than the one observed, assuming that the null hypothesis is true. In this case, the null hypothesis is that there is no difference between the features of the group of fake reviews and the group of real reviews. If the p-value is less than 0.05, we reject the null hypothesis and conclude that there is a significant difference between the features of the two groups. If the p-value is greater than or equal to 0.05, we fail to reject the null hypothesis and conclude that there is not enough evidence to conclude that there is a significant difference. Note that we can simply reject or not reject the null hypothesis - we can never conclude that we accept the alternate hypothesis.

6. **Making a decision**: Based on the p-value and the level of significance, a decision is made on whether to reject the null hypothesis or not. If the p-value is less than 0.05, we reject the null hypothesis and conclude that there is a significant difference between the features of the group of fake reviews and the group of real reviews. If the p-value is greater than or equal to 0.05, we fail to reject the null hypothesis and conclude that there is not enough evidence to conclude that there is a significant difference.

7. **Drawing conclusions**: The final step is to draw conclusions based on the results of the hypothesis test and to determine whether the evidence supports the alternative hypothesis or not. In this case, if we reject the null hypothesis, we conclude that there is a significant difference between the features of the group of fake reviews and the group of real reviews. If we fail to reject the null hypothesis, we conclude that there is not enough evidence to conclude that there is a significant difference between the features of the two groups.

Overall, hypothesis testing is a powerful statistical method used to test claims about populations using sample data. By following these given steps, we can make informed decisions based on data and draw valid conclusions about whether there is a significant difference between the features of the group of fake reviews and the group of real reviews.

Note that here we have used real and fake reviews as an example. In theory, this experiment can be repeated for any two groups and any features. In the chapters to come, we will see several examples of varied datasets (containing images, text, video, and malware). We will also collect different kinds of features (image feature vectors, linguistic features, and API call sequences). While we will not conduct hypothesis testing every time, you are highly encouraged to do so in order to strengthen your understanding of the concepts.

Now that we have defined clearly the steps involved in hypothesis testing, let us look at the most crucial part of it: the actual tests being conducted.

T-tests

T-tests are a statistical method used to compare the means of two groups and determine whether there is a statistically significant difference between them. The t-test is a hypothesis test that is based on the t-distribution, which is similar to the normal distribution but with a heavier tail.

There are two main types of t-tests:

- **Independent samples t-test**: This test is used when we want to compare the means of two independent groups, such as a treatment group and a control group, or a group of men and a group of women. The null hypothesis is that there is no difference in the means of the two groups.

- **Paired samples t-test**: This test is used when we want to compare the means of two related groups, such as before-and-after measurements for the same individuals, or measurements taken from two matched groups. The null hypothesis is that there is no difference between the means of the two related groups.

The t-test calculates a t-value, which is a measure of how different the means of the two groups are, relative to the variability within each group. The t-value is calculated by dividing the difference between the means of the two groups by the standard error of the difference. The standard error of the difference takes into account both the sample sizes and the variances of the two groups.

T-tests can be one-sided or two-sided:

- **One-sided t-test**: A one-sided t-test, also known as a directional t-test, is used when we have a specific hypothesis about the direction of the difference between the means of the two groups. For example, we might hypothesize that the mean of one group is greater than the mean of the other group. In this case, we would use a one-sided t-test to test this hypothesis. The null hypothesis for a one-sided t-test is that there is no difference between the means of the two groups in the hypothesized direction.

- **Two-sided t-test**: A two-sided t-test, also known as a non-directional t-test, is used when we do not have a specific hypothesis about the direction of the difference between the means of the two groups. For example, we might simply want to test whether the means of the two groups are different. In this case, we would use a two-sided t-test. The null hypothesis for a two-sided t-test is that there is no difference between the means of the two groups.

Once the t-value is calculated, we can determine the p-value, which is the probability of obtaining a t-value as extreme or more extreme than the one observed, assuming that the null hypothesis is true. If the p-value is less than the level of significance (usually 0.05), we reject the null hypothesis and conclude that there is a statistically significant difference between the means of the two groups.

T-tests are commonly used in many fields, such as psychology, biology, economics, and engineering, to compare the means of two groups and make statistical inferences. Note that in t-tests, we always decide on whether the null hypothesis is to be rejected based on the p-value. If the p-value is higher than our chosen threshold, we never say that we *accept* the alternate hypothesis; we simply say that we *failed to reject* the null hypothesis.

Conducting t-tests

We will now see how the t-test is actually conducted in Python and what information it gives us. Let us use the t-test to compare the distribution of the *review text length* feature in the fake and real reviews group. Therefore, our hypotheses are as follows:

- The null hypothesis (H0) is that the mean review length is the same for real and fake reviews

- The alternate hypothesis (H1) is that the mean review length between the two groups is different

Note that this is an unpaired t-test as samples in both groups are independent. This is also a two-tailed t-test as our hypothesis is not directional. Let us first plot the distribution of the review length for the two groups. We will obtain an array representing the review lengths within the two groups separately, and then plot their histograms on the same grid:

```
import matplotlib.pyplot as plt

# Separate real and fake reviews
fake_reviews = reviews_df[reviews_df['FRAUD_LABEL'] == 1]['Review_
Text_Length'].values
```

```
real_reviews = reviews_df[reviews_df['FRAUD_LABEL'] == 0]['Review_
Text_Length'].values

# Plot the two histograms
bins = np.linspace(0, 500, 500)
plt.hist(fake_reviews, bins, alpha=0.5, label='Fake')
plt.hist(real_reviews, bins, alpha=0.5, label='Real')

# Label the plot
plt.xlabel("Review Length")
plt.ylabel("# Reviews")
plt.legend()

# Display the plot
plt.show()
```

Here is the output:

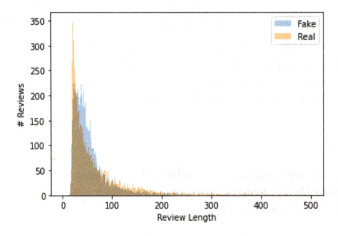

Figure 4.6 – Review length distribution across real and fake reviews

So we can see that there are some distributional differences between the two groups. The real reviews have a sharp peak at the lower values of the feature while the fake reviews have several intermediate peaks. Now we will do a t-test to see whether the difference is statistically significant. We will use the scipy module in Python to do this.

scipy is a powerful and widely used scientific computing library for the Python programming language. It provides a broad range of functionality for scientific and technical computing, including optimization, integration, interpolation, linear algebra, signal and image processing, and more. It is built on top of the NumPy library, which provides support for large, multi-dimensional arrays and matrices, and extends its capabilities to higher-level mathematical functions.

One of the key strengths of `scipy` is its sub-modules, which provide specialized functionality for different areas of scientific computing:

- The `optimization` sub-module provides functions for finding the minimum or maximum of a function, root-finding, curve-fitting, and more

- The `integration` sub-module provides functions for numerical integration and solving differential equations

- The `interpolate` sub-module provides functions for interpolating and smoothing data

- The `linalg` sub-module provides functions for linear algebra, including matrix decompositions, solving linear systems of equations, and more

- The `signal` sub-module provides functions for signal processing, including filtering, Fourier transforms, and more

- The `sparse` sub-module provides sparse matrix implementations and related operations

The following code will conduct the t-test and print out some statistics from it:

```
from scipy.stats import ttest_ind

# Conduct t-test
t_stat, p_value = ttest_ind(fake_reviews, real_reviews)

# Print group means
print("Mean in Fake Reviews: ", np.mean(fake_reviews))
print("Mean in Real Reviews: ", np.mean(real_reviews))

# Print t-test statistics
print("T-statistic value: ", t_stat)
print("P-Value: ", p_value)
```

After running this, you will see the following output:

```
Mean in Fake Reviews:   59.62095238095238
Mean in Real Reviews:   80.63780952380952
T-statistic value:   -17.56360707600074
P-Value:   1.4478425823590511e-68
```

This tells us a couple of things:

- The mean review length in the fake reviews is 59.62 while in the real reviews, it is 80.63. Therefore, the average fake review is nearly 21 words shorter than the average real review.

- Our p-value comes out to be 1.447e-68, which is of the order of 10^{-68} and is therefore very small. As this value is much smaller than 0.05, we can conclude that the difference we saw previously was statistically significant.

- As p < 0.05, this implies that the null hypothesis can be rejected. Therefore, we can conclude that the mean review length differs between real and fake reviews and this difference is statistically significant.

We can repeat this exercise for any feature of our choice. For example, here is how we do it for the number of words misspelled. The code is more or less the same, but note the change in the bins. The bins depend on the range of the feature under consideration. To plot the data, we use the following code:

```
import matplotlib.pyplot as plt


# Separate real and fake reviews
fake_reviews = reviews_df[reviews_df['FRAUD_LABEL'] == 1]['Num_
Misspelling'].values
real_reviews = reviews_df[reviews_df['FRAUD_LABEL'] == 0]['Num_
Misspelling'].values


# Plot the two histograms
bins = np.linspace(0, 50, 50)
plt.hist(fake_reviews, bins, alpha=0.5, label='Fake')
plt.hist(real_reviews, bins, alpha=0.5, label='Real')


# Label the plot
plt.xlabel("Review Length")
plt.ylabel("# Misspelt Words")
plt.legend()


# Display the plot
plt.show()
```

This shows us the distribution as follows:

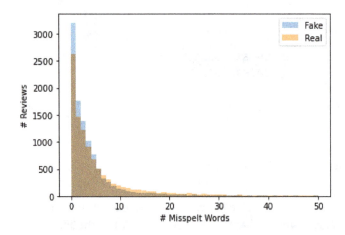

Figure 4.7 – Distribution of misspelled words across real and fake reviews

The distribution looks pretty similar. Let's do the t-test:

```
from scipy.stats import ttest_ind

# Conduct t-test
t_stat, p_value = ttest_ind(fake_reviews, real_reviews)

# Print group means
print("Mean in Fake Reviews: ", np.mean(fake_reviews))
print("Mean in Real Reviews: ", np.mean(real_reviews))

# Print t-test statistics
print("T-statistic value: ", t_stat)
print("P-Value: ", p_value)
```

Here's what we see:

```
Mean in Fake Reviews:  4.716952380952381
Mean in Real Reviews:  7.844952380952381
T-statistic value:  -18.8858003626682
P-Value:  6.730184744054038e-79
```

So this shows us that there are statistically significant differences between these two features as well.

Hypothesis testing is a useful tool that helps us examine signals that differentiate between groups, and the statistical significance of those signals. It helps data scientists and statisticians make claims about the data and give them mathematical backing.

A note on ANOVA tests

In the previous section, we learned how t-tests can help compare the means of two dependent or independent groups. This is helpful when you have a clean binary class problem. However, there often may be multiple classes. For example, there could be real reviews, bot reviews, incentivized reviews, and crowdsourced reviews all labeled separately. In such cases, we compare the means using an **Analysis of Variance (ANOVA)** test.

ANOVA is a statistical method used to test the hypothesis that there is no significant difference between the means of two or more groups. ANOVA compares the variation within groups to the variation between groups to determine whether there is a significant difference in means. There are several types of ANOVA, but the most common is one-way ANOVA. One-way ANOVA is used to compare the means of two or more independent groups. For example, you might use one-way ANOVA to compare the average test scores of students in three different classes.

The basic idea behind ANOVA is to partition the total variation in the data into two sources: variation due to differences between groups and variation due to differences within groups. The ratio of these two sources of variation is then used to determine whether the means of the groups are significantly different. The results of ANOVA are typically reported as an F-statistic, which is a measure of the ratio of the between-group variation to the within-group variation. If the F-statistic is large enough (that is, if the between-group variation is much larger than the within-group variation), then the null hypothesis of no difference between the groups is rejected, and it can be concluded that there is a significant difference between the means of the groups.

As we only have two classes here, we will not be implementing ANOVA or seeing it in practice. However, the `scipy` library contains implementations for ANOVA just like it does for t-tests, and I encourage you to go through it.

Now that we have seen how to examine feature differences, let us turn toward modeling our data with the most fundamental machine learning algorithm – linear regression.

Modeling fake reviews with regression

In this section, we will use the features we examined to attempt to model our data with linear regression.

Ordinary Least Squares regression

Ordinary Least Squares (OLS) linear regression is a statistical method used to model the relationship between a dependent variable and one or more independent variables. The goal of OLS is to find the linear function that best fits the data by minimizing the sum of squared errors between the observed values and the predicted values of the dependent variable.

The linear function is typically expressed as:

$$Y = \beta_0 + \beta_1 X_1 + \beta_2 X_2 + ... + \beta_n X_n + \varepsilon$$

where Y is the dependent variable, X_1, X_2, ..., X_n are the independent variables, β_0, β_1, β_2, ..., β_n are the coefficients (or parameters) that measure the effect of each independent variable on the dependent variable, and ε is the error term (or residual) that captures the part of the dependent variable that is not explained by the independent variables.

The OLS method estimates the coefficients by finding the values that minimize the sum of squared errors:

$$Loss = \Sigma \left(y_i - \hat{y}_i \right)^2$$

where y_i is the observed value of the dependent variable, $\hat{y}_i$ is the predicted value of the dependent variable based on the linear function, and the sum is taken over all the observations in the sample.

Once the values of the coefficients have been estimated, they can be plugged back into the equation to compute the predicted output value based on the defined equation. In this section, we will explore how linear regression can be used to detect whether a given review is fake or not.

OLS assumptions

OLS regression makes several assumptions about the data, and violating these assumptions can lead to biased or inefficient estimates of the regression coefficients. Here are the main assumptions of OLS regression:

- **Linearity**: The relationship between the dependent variable and the independent variables is linear. This means that the effect of each independent variable on the dependent variable is constant across the range of values of the independent variable. In the context of fake review detection, this assumption would mean that the relationship between the textual features (such as sentiment scores or word frequencies) and the likelihood of a review being fake is linear.

- **Independence**: The observations are independent of each other, meaning that the value of one observation does not depend on the value of another observation. For example, in fake review detection, this assumption would mean that the reviews are not systematically related to each other, such as being written by the same person or being about the same product.

- **Homoscedasticity**: The variance of the errors (residuals) is constant across all values of the independent variables. This means that the spread of the residuals does not change as the values of the independent variables change. In the context of fake review detection, this assumption would mean that the variability in the likelihood of a review being fake is the same for all levels of the textual features.

- **Normality**: The errors are normally distributed with a mean of zero. This means that the distribution of the residuals is symmetrical around 0 and follows a bell-shaped curve. In the context of fake review detection, this assumption would mean that the errors in the predicted probabilities of a review being fake are normally distributed.

- **No multicollinearity**: There is no perfect correlation between any pair of independent variables. This means that the independent variables are not redundant or highly correlated with each other. In the context of fake review detection, this assumption would mean that the textual features are not too similar to each other, as this could lead to multicollinearity and make it difficult to identify which features are driving the probability of a review being fake.

To ensure that these assumptions are met in practice, it is important to carefully preprocess the data and to check for violations of these assumptions during and after the modeling process. For example, in the context of fake review detection, we might check for independence by removing reviews written by the same person or about the same product, and we might check for normality by plotting the residuals and checking for a normal distribution.

Interpreting OLS regression

In this section, we will look at the metrics we use to evaluate the OLS regression model, namely the **R-squared (R^2)**, F-statistic, and regression coefficients.

R^2

In OLS linear regression, R^2 is a statistical measure that represents the proportion of variance in the dependent variable that can be explained by the independent variables in the model. It provides a measure of how well the regression line fits the data. R^2 takes values between 0 and 1, with 0 indicating that none of the variation in the dependent variable is explained by the independent variables, and 1 indicating that all of the variation in the dependent variable is explained by the independent variables.

Adjusted R^2 is a modified version of the R^2 value in OLS linear regression that takes into account the number of independent variables in the model. It is calculated using the following formula:

$$Adj(R^2) = 1 - \frac{(1 - R^2)(n - 1)}{n - k - 1}$$

Here, n is the number of observations in the sample and k is the number of independent variables in the model. The adjusted R^2 value provides a more conservative estimate of the goodness of fit of the model than the R^2 value. Unlike the R^2 value, which increases as the number of independent variables in the model increases, the adjusted R^2 value penalizes models that have additional independent variables that do not significantly improve the fit of the model.

The F-statistic

The F-statistic is a statistical test used in OLS regression to determine whether the overall regression model is statistically significant or not. It is calculated by comparing the variance explained by the regression model to the variance not explained by the model. In particular, the F-statistic measures the ratio of the **mean square of the regression (MSR)** to the **mean squared error (MSE)**. The MSR represents the variation in the dependent variable that is explained by the independent variables in the model, while the MSE represents the unexplained variation in the dependent variable that is not accounted for by the model.

A high F-statistic with a low associated p-value suggests that the regression model as a whole is statistically significant and that at least one of the independent variables in the model is related to the dependent variable. In contrast, a low F-statistic with a high associated p-value suggests that the model is not statistically significant and that the independent variables do not significantly explain the variation in the dependent variable.

Regression coefficients

Regression coefficients represent the change in the dependent variable for a one-unit change in the independent variable, holding all other independent variables constant. More specifically, in linear regression, the coefficients indicate the slope of the regression line, which represents the relationship between the independent variable and the dependent variable. A positive coefficient means that as the independent variable increases, the dependent variable also increases. A negative coefficient means that as the independent variable increases, the dependent variable decreases. The size of the coefficient indicates the magnitude of the effect of the independent variable on the dependent variable. Larger coefficients indicate a stronger relationship between the independent variable and the dependent variable, while smaller coefficients indicate a weaker relationship.

It is important to note that the interpretation of coefficients may be affected by the scaling of the variables. For example, if one independent variable is measured in dollars and another independent variable is measured in percentages, the coefficients cannot be directly compared, as they are not on the same scale. Additionally, coefficients may also be affected by collinearity among the independent variables, which can make it difficult to distinguish the unique effect of each independent variable on the dependent variable. Therefore, when interpreting regression coefficients, it is important to consider the scale of the variables, the context of the study, and potential confounding factors.

Implementing OLS regression

We will attempt to model the fake reviews problem with an OLS regression model using the `statsmodels` package, which is a comprehensive Python package that provides a wide range of statistical tools and models for data analysis. It is designed to be used in conjunction with other scientific computing libraries such as `NumPy`, `sciPy`, and `pandas`. One of the key features of `statsmodels` is its ability to estimate statistical models for different types of data. For example, it provides a range of models for linear regression, generalized linear models, time-series analysis, and multilevel models. These models can be used for a variety of tasks such as prediction, classification, and inference.

In addition to its model estimation capabilities, `statsmodels` also includes a range of statistical tests and diagnostics. These tools can be used to assess the quality of a model and determine whether its assumptions are being violated. Some of the statistical tests provided by `statsmodels` include hypothesis tests, goodness-of-fit tests, and tests for stationarity and cointegration. `statsmodels` also includes a range of visualization tools for exploring data and model results. These tools can help users gain insights into their data and communicate their findings effectively. Some of the visualization tools provided by `statsmodels` include scatter plots, line plots, histograms, and QQ plots.

In order to implement OLS regression, you specify a formula that indicates the dependent variable and the independent variables we want to model it on. Here is how it can be done in the `statsmodels` library:

```
import statsmodels.formula.api as smf

model = smf.ols(formula = 'FRAUD_LABEL ~ Review_Text_Length + Num_
Misspelling + Avg_Word_Len',
                data = review_df).fit()

print(model.summary())
```

The output is as follows:

```
                          OLS Regression Results
============================================================================
Dep. Variable:             FRAUD_LABEL   R-squared:                    0.017
Model:                             OLS   Adj. R-squared:               0.017
Method:                  Least Squares   F-statistic:                  121.7
Date:                 Thu, 16 Mar 2023   Prob (F-statistic):        4.00e-78
Time:                         15:49:06   Log-Likelihood:              -15061.
No. Observations:                21000   AIC:                       3.013e+04
Df Residuals:                    20996   BIC:                       3.016e+04
Df Model:                            3
Covariance Type:             nonrobust
============================================================================
                       coef    std err          t      P>|t|     [0.025      0.975]
----------------------------------------------------------------------------
Intercept            0.5280      0.007     81.062      0.000      0.515       0.541
Review_Text_Length  -0.0001   8.46e-05     -1.742      0.081     -0.000    1.84e-05
Num_Misspelling     -0.0044      0.001     -7.219      0.000     -0.006      -0.003
Avg_Word_Len         0.0023      0.001      2.209      0.027      0.000       0.004
============================================================================
Omnibus:                     74064.046   Durbin-Watson:                0.033
Prob(Omnibus):                   0.000   Jarque-Bera (JB):          3321.541
Skew:                           -0.011   Prob(JB):                      0.00
Kurtosis:                        1.052   Cond. No.                      216.
============================================================================

Notes:
[1] Standard Errors assume that the covariance matrix of the errors is correctly specified.
```

Figure 4.8 – OLS regression results

Let us interpret these results:

- The R-squared value of 0.017 indicates that only 1.7% of the variance in the dependent variable (**FRAUD_LABEL**) is explained by the independent variables (**Review_Text_Length**, **Num_Misspelling**, and **Avg_Word_Len**). This means that the model is not very effective in predicting fraud based on the provided independent variables. In general, a higher R-squared value is better.

- The coefficients for the independent variables give an indication of how each variable affects the dependent variable. The coefficient for **Review_Text_Length** is negative, but not statistically significant ($P>|t| = 0.081$). This suggests that there may be a weak negative relationship between the length of the review text and the likelihood of fraud, but this relationship is not strong enough to be statistically significant.

- The coefficient for **Num_Misspelling** is negative and statistically significant (P>|t| = 0.000). This suggests that there is a strong negative relationship between the number of misspellings in a review and the likelihood of fraud. Specifically, for each additional misspelling, the likelihood of fraud decreases by 0.0044.

- The coefficient for **Avg_Word_Len** is positive and statistically significant (P>|t| = 0.027). This suggests that there is a weak positive relationship between the average word length in a review and the likelihood of fraud. Specifically, for each additional unit of average word length, the likelihood of fraud increases by 0.0023.

- The Omnibus test shows that the residuals are not normally distributed (Prob(Omnibus) < 0.05), which suggests that the normality assumption of the model may not hold. The Durbin-Watson value of 0.033 indicates that there may be autocorrelation in the residuals, while the Jarque-Bera test suggests that there may be some small deviation from normality in the residuals. These results should be investigated further to determine whether they are problematic for the analysis.

Overall, the regression model suggests that the number of misspellings in a review is the most important predictor of fraud, while the length of the review text and the average word length are weaker predictors. However, the R^2 value is low, indicating that the model is not a strong predictor of fraud in general.

In our previous example, we used only continuous features that we had derived. However, it is possible to use categorical features as well. The formula that we pass to the OLS regression function can be amended as follows:

```
import statsmodels.formula.api as smf

model = smf.ols(formula = """FRAUD_LABEL ~ Review_Text_Length + Num_
Misspelling + Avg_Word_Len + C(RATING) + C(VERIFIED_PURCHASE)""", data
= reviews_df).fit()
print(model.summary())
```

And you see the following result:

```
                              OLS Regression Results
===============================================================================
Dep. Variable:            FRAUD_LABEL   R-squared:                      0.359
Model:                            OLS   Adj. R-squared:                 0.359
Method:                 Least Squares   F-statistic:                    1470.
Date:                Fri, 17 Mar 2023   Prob (F-statistic):             0.00
Time:                        15:29:31   Log-Likelihood:                -10570.
No. Observations:               21000   AIC:                         2.116e+04
Df Residuals:                   20991   BIC:                         2.123e+04
Df Model:                           8
Covariance Type:            nonrobust
===============================================================================
                             coef    std err          t      P>|t|      [0.025      0.975]
-------------------------------------------------------------------------------
Intercept                  0.8513      0.011     77.712      0.000       0.830       0.873
C(RATING)[T.2]            -0.0086      0.015     -0.569      0.569      -0.038       0.021
C(RATING)[T.3]            -0.0316      0.013     -2.373      0.018      -0.058      -0.005
C(RATING)[T.4]            -0.0086      0.011     -0.751      0.453      -0.031       0.014
C(RATING)[T.5]             0.0520      0.010      5.085      0.000       0.032       0.072
C(VERIFIED_PURCHASE)[T.Y] -0.5951      0.006   -105.806      0.000      -0.606      -0.584
Review_Text_Length        -0.0003   6.84e-05     -4.546      0.000      -0.000      -0.000
Num_Misspelling           -0.0051      0.000    -10.362      0.000      -0.006      -0.004
Avg_Word_Len               0.0020      0.001      2.406      0.016       0.000       0.004
===============================================================================
Omnibus:                      205.758   Durbin-Watson:                   0.696
Prob(Omnibus):                  0.000   Jarque-Bera (JB):              211.719
Skew:                           0.246   Prob(JB):                     1.06e-46
Kurtosis:                       2.997   Cond. No.                        931.
===============================================================================

Notes:
[1] Standard Errors assume that the covariance matrix of the errors is correctly specified.
```

Figure 4.9 – OLS regression with categorical features

From this output, we can see that the R^2 is now 0.359, which indicates that the independent variables explain about 36% of the variation in the dependent variable. This is an improvement over our previous model, where the independent variables could explain less than 2% of the variance in the dependent variables.

While the interpretation for the other variables (review text length, average word length, and number of misspellings) remains the same, we have also added categorical variables to our analysis. In this OLS regression model, the coefficients for **RATING** and **VERIFIED_PURCHASE** correspond to the estimated difference in the mean value of the dependent variable, **FRAUD_LABEL**, for the corresponding category and the reference category.

For **RATING**, the reference category is assumed to be the lowest rating (1). The estimated coefficients for **RATING** suggest that, compared to the lowest rating, the mean value of **FRAUD_LABEL** is as follows:

- 0.0086 units lower for a rating of 2 (not significant, $p > 0.05$)

- 0.0316 units lower for a rating of 3 (significant, $p < 0.05$)

- 0.0086 units lower for a rating of 4 (not significant, $p > 0.05$)

- 0.0520 units higher for a rating of 5 (significant, $p < 0.05$)

For **VERIFIED_PURCHASE**, the reference category is assumed to be a non-verified purchase (N). The estimated coefficient for **VERIFIED_PURCHASE** indicates that the mean value of **FRAUD_LABEL** is 0.5951 units lower for verified purchases (significant, $p < 0.05$) compared to non-verified purchases. This suggests that verified purchases are associated with a lower probability of fraud.

It's worth noting that the coefficients for categorical variables in this model are estimated relative to the reference category. Therefore, if you were to change the reference category for **RATING** or **VERIFIED_PURCHASE**, the estimated coefficients would change, but the overall model fit and significance levels would remain the same.

Summary

In this chapter, we examined the problem of fake reviews on e-commerce platforms through the lens of statistical and machine learning models. We began by understanding the review ecosystem and the nature of fake reviews, including their evolution over time. We then explored a dataset of fake reviews and conducted statistical tests to determine whether they show characteristics significantly different from genuine reviews. Finally, we modeled the review integrity using OLS regression and examined how various factors affect the likelihood that a review is fake.

This chapter introduced you to the foundations of data science, including exploratory data analysis, statistics, and the beginnings of machine learning.

In the next chapter, we will discuss techniques for detecting deepfakes, which plague the internet and social media today.

5

Detecting Deepfakes

In recent times, the problem of deepfakes has become prevalent on the internet. Easily accessible technology allows attackers to create images of people who have never existed, through the magic of deep neural networks! These images can be used to enhance fraudulent or bot accounts to provide an illusion of being a real person. As if deepfake images were not enough, deepfake videos are just as easy to create. These videos allow attackers to either morph someone's face onto a different person in an existing video, or craft a video clip in which a person says something. Deepfakes are a hot research topic and have far-reaching impacts. Abuse of deepfake technology can result in misinformation, identity theft, sexual harassment, and even political crises.

This chapter will focus on machine learning methods to detect deepfakes. First, we will understand the theory behind deepfakes, how they are created, and what their impact can be. We will then cover two approaches to detecting deepfake images – vanilla approaches using standard machine learning models, followed by some advanced methods. Finally, as deepfake videos are major drivers of misinformation and have the most scope for exploitation, the last part of the chapter will focus on detecting deepfake videos.

In this chapter, we will cover the following main topics:

- All about deepfakes
- Detecting fake images
- Detecting fake videos

By the end of this chapter, you will have an in-depth understanding of deepfakes, the technology behind them, and how they can be detected.

Technical requirements

You can find the code files for this chapter on GitHub at https://github.com/ PacktPublishing/10-Machine-Learning-Blueprints-You-Should-Know-for-Cybersecurity/tree/main/Chapter%205.

All about deepfakes

The word *deepfake* is a combination of two words – *deep learning* and *fake*. Put simply, deepfakes are fake media created using deep learning technology. In the past decade, there have been significant advances in machine learning and generative models – models that create content instead of merely classifying it. These models (such as **Generative Adversarial Networks** (**GANs**)) can synthesize images that look real – even of human faces!

Deepfake technology is readily accessible to attackers and malicious actors today. It requires no sophistication or technical skills. As an experiment, head over to the website `thispersondoesnotexist.com`. This website allows you to generate images of people – people who have never existed!

For example, the people in the following figure are not real. They are deepfakes that have been generated by `thispersondoesnotexist.com`, and it only took a few seconds!

Figure 5.1 – Deepfakes generated from a website

Amazing, isn't it? Let us now understand what makes generating these images possible, and the role machine learning has to play in it.

A foray into GANs

Let us now look at GANs, the technology that makes deepfakes possible.

GANs are deep learning models that use neural networks to synthesize data rather than merely classify it. The name *adversarial* comes from the architectural design of these models; a GAN architecture consists of two neural networks, and we can force them into a cat-and-mouse game to generate synthetic media that can be passed off as real.

GAN architecture

A GAN consists of two main parts – the generator and the discriminator. Both models are deep neural networks. Synthetic images are generated by plotting these networks against each other:

- **Generator**: The generator is the model that learns to generate real-looking data. It takes a fixed-length random vector as input and generates an output in the target domain, such as an image. The random vector is sampled randomly from a Gaussian distribution (that is, a standard normal distribution, where most observations cluster around the mean, and the further away an observation is from the mean, the lower the probability of it occurring is).

- **Discriminator**: The discriminator is a traditional machine learning model. It takes in data samples and classifies them as real or fake. Positive examples (those labeled as real) come from a training dataset, and negative examples (those labeled as fake) come from the generator.

GAN working

Generating images is an unsupervised machine learning task and classifying the images is a supervised one. By training both the generator and discriminator jointly, we refine both and are able to obtain a generator that can generate samples good enough to fool the discriminator.

In joint training, the generator first generates a batch of sample data. These are treated as negative samples and augmented with images from a training dataset as positive samples. Together, these are used to fine-tune the discriminator. The discriminator model is updated to change parameters based on this data. Additionally, the discriminator loss (i.e., how well the generated images fooled the discriminator) is fed back to the generator. This loss is used to update the generator to generate better data.

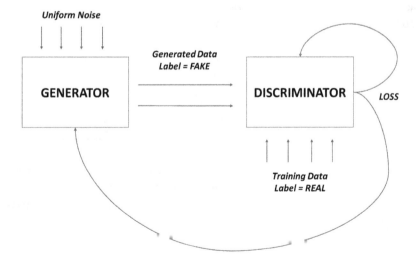

Figure 5.2 – How a GAN model works

At first, the generator produces random data that is clearly noisy and of poor quality. The discriminator learns to easily discern between real and fake examples. As the training progresses, the generator starts to get better as it leverages the signal from the discriminator, changing its generation parameters accordingly. In the end, in an ideal situation, the generator will have been so well trained that the generator loss decreases and the discriminator loss begins increasing, indicating that the discriminator can no longer effectively distinguish between actual data and generated data.

How are deepfakes created?

Along with being a portmanteau of the words *deep learning* and *fake*, the word *deepfake* has another origin. The very first deepfake video was created by a Reddit user by the name of `r/deepfakes`. This user used the open source implementation provided by Google to swap the face of several actresses into pornographic videos. Much of the amateur deepfakes in the wild today build upon this code to generate deepfakes.

In general, any deepfake creation entails the following four steps:

1. Analyzing the source image, identifying the area where the face is located, and cropping the image to that area.

2. Computing features that are typically representations of the cropped image in a latent low-dimensional space, thus encoding the image into a feature vector.

3. Modifying the generated feature vector based on certain signals, such as the destination image.

4. Decoding the modified vector and blending the image into the destination frame.

Most deepfake generation methods rely on neural networks and, in particular, the encoder-decoder architecture. The encoder transforms an image into a lower dimensional subspace and maps it to a latent vector (similar to the context vectors we described when discussing transformers). This latent vector captures features about the person in the picture, such as color, expression, facial structure, and body posture. The decoder does the reverse mapping and converts the latent representation into the target image. Adding a GAN into the mix leads to a much more robust and powerful deepfake generator.

The first commercial application of deepfakes started with the development of FakeApp in January 2018. This is a desktop application that allows users to create videos with faces swapped for other faces. It is based on autoencoder architecture and consists of two encoder-decoder pairs. One encoder-decoder pair is trained on the images of the source image, and the other is trained on the images of the target. However, both encoders have shared common weights; in simpler terms, the same encoder is used in both autoencoders. This means that the latent vector representation for both images is in the same context, hence representing similar features. At the end of the training period, the common encoder will have learned to identify common features in both faces.

Let's say A is the source image (the original image of our victim) and B is the target image (the one where we want to insert A). The high-level process to generate deepfakes using this methodology is as follows:

1. Obtain multiple images of A and B using data augmentation techniques and image transformations so that the same picture from multiple viewpoints is considered.

2. Train the first autoencoder model on images of A.

3. Extract the encoder from the first autoencoder. Use this to train a second autoencoder model on images of B.

4. At the end of training, we have two autoencoders with a shared encoder that can identify common features and characteristics in the images of A and B.

5. To produce the deepfake, pass image A through the common encoder and obtain the latent representation.

6. Use the decoder from the second autoencoder (i.e., the decoder for images of B) to decode this back into the deepfake image. The following shows a diagram of this process:

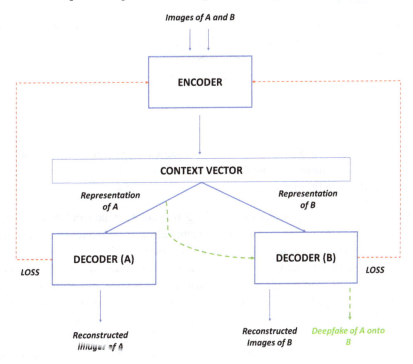

Figure 5.3 – Deepfake creation methodology

FakeApp has been widely used and inspired many other open source implementations, such as DeepFaceLab, FaceSwap, and Dfaker.

The social impact of deepfakes

While deepfake technology is certainly revolutionary in the field of machine learning and deep learning in particular, it has a far-reaching societal impact. Since their inception, deepfakes have been used for both benign and malicious purposes. We will now discuss both briefly. Understanding the full impact of deepfakes is essential to their study, especially for machine learning practitioners in the cybersecurity industry.

Benign

Not all deepfakes are bad. There are some very good reasons why the use of deepfakes may be warranted and, at times, beneficial. Deepfakes have shown to be of great utility in some fields:

- **Entertainment**: Deepfake technology is now accessible to everyone. Powerful pre-trained models allow the exposure of endpoints through apps on smartphones, where they have been widely used to create entertaining videos. Such deepfakes include comedy videos where cartoons say certain dialogue, images where the faces of two people are swapped, or filters that generate human faces morphed into animals.

- **Resurrection**: An innovative use case of deepfakes has been reviving the deceased. Deepfakes can portray deceased people saying certain things, in their own voice! This is like magic, especially for someone who has no videographic memories of themselves left in this world. Deepfakes have also been used to create images that are able to portray how someone would have looked in a few years' or decades' time.

Malicious

Unfortunately, however, every coin has two sides. The same deepfake technology that can power entertainment and enable resurrected digital personas can be used maliciously by attackers. Here are a few attack vectors that practitioners in this space should be aware of:

- **Misinformation**: This has probably been the biggest use of deepfakes and the most challenging one to solve. Because deepfakes allow us to create videos of someone saying things they did not, this has been used to create videos that spread fake news. For example, a video featuring a surgeon general saying that vaccines are harmful and cause autism would certainly provoke widespread fear and lead people to believe that it is true. Malicious entities can create and disseminate such deepfakes to further their own causes. This can also lead to political crises and loss of life. In 2022, a deepfake video featuring the Ukrainian president Volodymyr Zelenskyy was created by Russian groups, where the former was shown to accept defeat and ask soldiers to stand down – this was not the case, but the video spread like wildfire on social media before it was removed.

- **Fraud**: Traditionally, many applications depended on visual identification verification. After the COVID-19 pandemic, most of these transitioned to verifying identities through documents and video online. Deepfakes have tremendous potential to be used for identity theft here; by crafting a deepfake video, you can pretend to be someone you are not and bypass automatic identity verification systems. In June 2022, the **Federal Bureau of Investigation** (**FBI**) published a public service announcement warning companies of deepfakes being used in online interviews and during remote work. Deepfakes can also be used to create sock puppet accounts on social media websites, GANs can be used to generate images, and deepfake videos can be used to enrich profiles with media (videos showing the person talking) so that they appear real.

- **Pornography**: The very first deepfake that was created was a pornographic movie. Revenge porn is an alarmingly growing application of deepfakes in today's world. By using images of a person and any pornographic clip as a base, it is possible to depict the said person in the pornographic clip. The naked eye may not be able to ascertain that the video is a spoof.

Detecting fake images

In the previous section, we looked at how deepfake images and videos can be generated. As the technology to do so is accessible to everyone, we also discussed the impact that this can have at multiple levels. Now, we will look at how fake images can be detected. This is an important problem to solve and has far-reaching impacts on social media and the internet in general.

A naive model to detect fake images

We know that machine learning has driven significant progress in the domain of image processing. Convolutional neural networks (CNNs) have surpassed prior image detectors and achieved accuracy even greater than that of humans. As a first step toward detecting deepfake images, we will treat the task as a simple binary classification and use standard deep learning image classification approaches.

The dataset

There are several publicly available datasets for deepfake detection. We will use the 140k Real and Fake Faces Dataset. This dataset is freely available to download from Kaggle (`https://www.kaggle.com/datasets/xhlulu/140k-real-and-fake-faces`).

As the name suggests, the dataset consists of 140,000 images. Half of these are real faces from Flickr, and the remaining half are fake ones generated from a GAN. The real faces have been collected by researchers from NVIDIA, and there is significant coverage of multiple ethnicities, age groups, and accessories in the images.

The dataset is fairly large (~4 GB). You will need to download the compressed dataset from Kaggle, unzip it, and store it locally. The root directory consists of three folders – one each for training, testing, and validation data.

The model

We are going to use a CNN model for classification. These neural networks specialize in understanding and classifying data in which the spatial representation of data matters. This makes it ideal for processing images, as images are fundamentally matrices of pixels arranged in a grid-like topology. A CNN typically has three main layers – convolution, pooling, and classification:

- **Convolution layer**: This is the fundamental building block of a CNN. This layer traverses through an image and generates a matrix known as an **activation map**. Each convolution multiplies some part of the image with a kernel matrix (the weights in the matrix are parameters that can be learned using gradient descent and backpropagation). The convolution matrix slides across the image row by row and performs the dot product. The convolution layer aggregates multiple neighborhoods of the image and produces a condensed output, as shown in *Figure 5.4*:

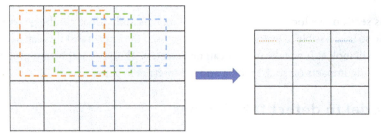

Figure 5.4 – The convolution layer

- **Pooling layer**: The pooling layer performs aggregations over the convolution output. It calculates a summary statistic over a neighborhood and replaces the neighborhood cells with the summary. The statistic can be the mean, max, median, or any other standard metric. Pooling reduces redundancy in the convolutional representation and reduces dimensions by summarizing over multiple elements, as shown in *Figure 5.5*:

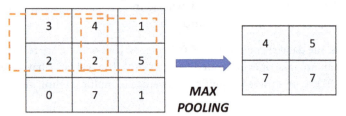

Figure 5.5 – The pooling layer

- **Fully connected layer**: The output from the pooling layer is flattened and converted into a one-dimensional vector. This is done simply by appending the rows to each other. This vector is now passed to a fully connected neural network, at the end of which is a **softmax layer**. This layer operates as a standard neural network, with weights being updated with gradient descent and backpropagation through multiple epochs. The softmax layer will output a probability distribution of the predicted class of the image, as shown in *Figure 5.6*:

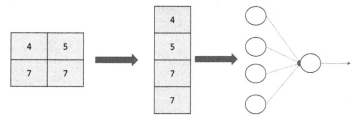

Figure 5.6 – Flattening and a fully connected layer

While training, data flows into the network and undergoes convolution and pooling. We can stack multiple convolution-pooling layers one after the other; the hope is that each layer will learn something new from the image and obtain more and more specific representations. The pooling output after the last convolution layer flows into the fully connected neural network (which can also consist of multiple layers). After the softmax output, a loss is calculated, which then flows back through the whole network. All layers, including the convolution layers, update their weights, using gradient descent to minimize this loss.

Putting it all together

We will now use a CNN model to detect deepfakes and run this experiment on the 140k dataset.

First, we import the required libraries:

```
# Data Processing
import numpy as np
import pandas as pd
import scipy.misc
from sklearn.datasets import load_files
import matplotlib.pyplot as plt
%matplotlib inline


# Deep Learning Libraries
from keras.models import Sequential, Model
from keras.layers import Input, Dense, Flatten, Dropout, Activation,
Lambda, Permute, Reshape
from keras.layers import Convolution2D, ZeroPadding2D, MaxPooling2D
from keras_vggface.vggface import VGGFace
from keras_vggface import utils
from keras.preprocessing import image
from keras.preprocessing.image import ImageDataGenerator
from keras.utils import np_utils
```

Recall that the first step in the data science pipeline is data preprocessing. We need to parse the images we downloaded and convert them into a form suitable for consumption by the CNN model. Fortunately, the `ImageDataGenerator` class in the `keras` library helps us do just that. We will use this library and define our training and test datasets:

```
training_data_path = ""
test_data_path = ""
batch_size = 64
print("Loading Train…")
training_data = ImageDataGenerator(rescale = 1./255.) .flow_from_
directory(
    training_data_path,
    target_size=(224, 224),
    batch_size=batch_size,
    class_mode='binary'
)
print("Loading Test…")
test_data = ImageDataGenerator(rescale = 1./255.)
.flow_from_directory(
    test_data_path,
    target_size=(224, 224),
    batch_size=batch_size,
    class_mode='binary'
)
```

After you run this, you should see an output somewhat like this:

```
Loading Train…
Found 100000 images belonging to 2 classes.
Loading Test…
Found 20000 images belonging to 2 classes.
```

Now that the data is preprocessed, we can define the actual model and specify the architecture of the CNN. Our model will consist of convolution, pooling, and the fully connected layer, as described earlier:

```
input_shape = (224,224,3)
epsilon=0.001
dropout = 0.1
model = Sequential()

# Convolution and Pooling -- 1
model.add(BatchNormalization(input_shape=input_shape))
model.add(Conv2D(filters=16, kernel_size=3, activation='relu',
padding='same'))
model.add(MaxPooling2D(pool_size=2))
```

```
model.add(BatchNormalization(epsilon=epsilon))

# Convolution and Pooling -- 2
model.add(Conv2D(filters=32, kernel_size=3, activation='relu',
padding='same'))
model.add(MaxPooling2D(pool_size=2))
model.add(BatchNormalization(epsilon=epsilon))
model.add(Dropout(dropout))

#Convolution and Pooling -- 3
model.add(Conv2D(filters=64, kernel_size=3, activation='relu',
padding='same'))
model.add(MaxPooling2D(pool_size=2))
model.add(BatchNormalization(epsilon=epsilon))
model.add(Dropout(dropout))

# Aggregation
model.add(GlobalAveragePooling2D())

# Fully Connected Layer
model.add(Dense(1, activation='sigmoid'))
```

Let us take a closer look at what we did here. First, we set some basic model parameters and defined an empty model using the `Sequential` class. This model will be our base, where we pile on different layers to build the whole architecture.

Then, we defined our **convolution layers**. Each convolution layer has a **max pooling layer**, followed by a **normalization layer**. The normalization layer simply normalizes and rescales the pooled output. This results in stable gradients and avoids exploding loss.

Then, we added an aggregation layer using the `GlobalAveragePooling2D` class. This will concatenate the previous output into a one-dimensional vector. Finally, we have a fully connected layer at the end with a softmax activation; this layer is responsible for the actual classification.

You can take a look at your model architecture by printing the summary:

```
model.summary()
```

Finally, we can train the model we defined on the data we preprocessed:

```
model.compile(loss='binary_crossentropy',optimizer=Adam(0.0001),
metrics=['accuracy'])

training_steps = 40000//batch_size
num_epochs = 10
history = model.fit_generator(
```

```
    training_data,
    epochs=num_epochs,
    steps_per_epoch = training_steps
)
```

This will take about 30 minutes to an hour to complete. Of course, the exact time will depend on your processor, scheduling, and system usage.

Once the model is trained, we can use it to make predictions for our test data and compare the predictions with the ground truth:

```
y_pred = model.predict(test_data)
y_actual = test_data.classes
```

Now, you can plot the confusion matrix, just like we did for the other models in the previous chapters.

Playing around with the model

In the previous section, we looked at how an end-to-end CNN model can be defined and how to train it. While doing so, we made several design choices that can potentially affect the performance of our model. As an experiment, you should explore these design choices and rerun the experiment with different parameters. We will not go through the analysis and hyperparameter tuning in detail, but we will discuss some of the parameters that can be tweaked. We will leave it up to you to test it out, as it is a good learning exercise. Here are some things that you could try out to play around with the model:

- **Convolutional layers**: In our model, we have three layers of convolution and max pooling. However, you can extend this to as many (or as few) layers as you want. What happens if you have 10 layers? What happens if you have only one? How does the resulting confusion matrix change?

- **Pooling layers**: We used max pooling in our model. However, as discussed in the CNN architecture, pooling can leverage any statistical aggregation. What happens if we use mean pooling or min pooling instead of max pooling? How does the resulting confusion matrix change?

- **Fully connected layers**: Our model has just one fully connected layer at the end. However, this does not have to be the case; you can have a full-fledged neural network with multiple layers. You should examine what happens if the number of layers is increased.

- **Dropout**: Note that each layer is followed by a dropout layer. Dropout is a technique used in neural networks to avoid overfitting. A certain fraction of the weights is randomly set to 0; these are considered to be "dropped out." Here, we have a dropout fraction of 0.1. What happens if it is increased? What happens if it is set to 0 (i.e., with no dropout at all)?

This completes our discussion of deepfake image detection. Next, we will see how deepfakes transitioned from images to videos, and we will look at methods for detecting them.

Detecting deepfake videos

As if deepfake images were not enough, deepfake videos are now revolutionizing the internet. From benign uses such as comedy and entertainment to malicious uses such as pornography and political unrest, deepfake videos are taking social media by storm. Because deepfakes appear so realistic, simply looking at a video with the naked eye does not provide any clues as to whether it is real or fake. As a machine learning practitioner working in the security field, it is essential to know how to develop models and techniques to identify deepfake videos.

A video can be thought of as an extension of an image. A video is multiple images arranged one after the other and viewed in quick succession. Each such image is known as a frame. By viewing the frames at a high speed (multiple frames per second), we see images moving.

Neural networks cannot directly process videos – there does not exist an appropriate method to encode images and convert them into a form suitable for consumption by machine learning models. Therefore, deepfake video detection involves deepfake image detection on each frame of the video. We will look at the succession of frames, examine how they evolve, and determine whether the transformations and movements are normal or similar to those expected in real videos.

In general, detecting video deepfakes follows the following pattern:

1. Parsing the video file and decomposing it into multiple frames.
2. Reading frames up to a maximum number of frames. If the number of frames is less than the maximum set, pad them with empty frames.
3. Detecting the faces in each frame.
4. Cropping the faces in each frame.
5. Training a model with a sequence of cropped faces (one face obtained per frame) as input and real/fake labels as output.

We will leverage these steps to detect fake videos in the next section.

Building deepfake detectors

In this section, we will look at how to build models to detect deepfake videos.

The dataset

We will use the dataset from the Kaggle Deepfake Detection Challenge (https://www.kaggle.com/competitions/deepfake-detection-challenge/overview). As part of the challenge, three datasets were provided – training, testing, and validation. With prizes worth $1 million, participants were required to submit code and output files, which were evaluated on a private test set.

The actual dataset is too large (around 0.5 TB) for us to feasibly download and process, given that most of you will have access to only limited compute power. We will use a subset of the data available in the `train_sample_videos.zip` and `test_videos.zip` files, which are around 4 GB altogether. You will need to download this data and save it locally so that you can run the experiments.

The model

In the previous section, we developed our own CNN model and trained it end to end for classification. In this section, we will explore a different technique based on the concept of *transfer learning*. In machine learning, transfer learning is a paradigm where parameters learned from one task can be applied to another task and improve the performance of the latter.

In transfer learning, we train a classification model on a base task. Once that model is trained, we can use it for another task in one of two ways:

- **Training the model again on new data from the second task**: By doing so, we will initialize the model weights as the ones produced after training on the first task. The hope is that the model (which will effectively be trained on both sets of data) will show a good performance on both tasks.

- **Using the model as a feature extractor**: The final layer of the model is typically a softmax layer. We can extract the pre-final layer and use the model without softmax to generate features for our new data. These features can be used to train a downstream classification model.

Note that the second task is generally a refined and specific version of the first task. For example, the first task can be sentiment classification, and the second task can be movie review classification. The hope is that training on the first task will help the model learn high-level signals for sentiment classifications (keywords indicating certain sentiments, such as `excellent`, `amazing`, `good`, and `terrible`). Fine-tuning in the second task will help it learn task-specific knowledge in addition to the high-level knowledge (movie-specific keywords such as `blockbuster` and `flop-show`). Another example is that the base task is image classification on animals, and the fine-tuning task is a classifier for cat and dog images.

We will use the second approach listed here. Instead of developing custom features, we will let a pre-trained model do all the work. The model we will use is the **InceptionV3 model**, which will extract features for every frame in the video. We will use a **Recurrent Neural Network (RNN)** to classify the video, based on the sequence of frames. This technique has been adapted from a submission to the challenge (`https://www.kaggle.com/code/krooz0/deep-fake-detection-on-images-and-videos/notebook`).

Putting it all together

First, we will import the necessary libraries:

```
# Libraries for Machine Learning
import tensorflow as tf
from tensorflow import keras
```

```
import pandas as pd
import numpy as np

# Helper Libraries
import imageio
import cv2
import os
```

We will then read the video file and label for each video into a DataFrame. The dataset contains a metadata file for each set that provides us with this information. Note that you may have to adjust the paths depending on where and how you stored the data locally:

```
train_meta_file = '../deepfake-detection-challenge/train_sample_
videos/metadata.json'
train_sample_metadata = pd.read_json(train_meta_file).T
```

Let's take a look at one of the DataFrames:

```
train_sample_metadata.head()
```

It should show you the list of video files along with their associated label (real or fake). We will now define a few helper functions. The first one is a cropping function that will take in a frame and crop it into a square:

```
def crop_image(frame):
    # Read dimensions
    y = frame.shape[0]
    x = frame.shape[1]

    # Calculate dimensions of cropped square
    min_dimension = min(x, y)
    start_x = (x/2) - (min_dimension/2)
    start_y = (y/2) - (min_dimension/2)

    # Pick only the part of the image to be retained
    cropped = frame[start_y : start_y + min_dim,
                    start_x : start_x + min_dim]
    return cropped
```

The second helper function will parse the video file and extract frames from it up to a specified maximum number of frames. We will resize each frame as (224, 224) for standardization and then crop it into a square. The output will be a sequence of uniformly sized and cropped frames:

```
def parse_video_into_frames(video_file_path):
    new_shape = (224, 224)
```

```
capture = cv2.VideoCapture(path)
frames = []
try:
    while True:
  # Read from the video stream
        ret, frame = capture.read()
        if not ret: # Have reached the end of frames
            break
  # Crop the frame and resize it
        frame = crop_image(frame)
        frame = cv2.resize(frame, new_shape)
        frame = frame[:, :, [2, 1, 0]]
    # Append it to a sequence
        frames.append(frame)
finally:
    capture.release()
return np.array(frames)
```

We will now define our feature extractor, which uses transfer learning and leverages the InceptionV3 model to extract features from a frame. Fortunately, the keras library provides a convenient interface for us to load pre-trained models:

```
inception_model = keras.applications.InceptionV3(
        weights="imagenet",
        include_top=False,
        pooling="avg",
        input_shape=(224,224, 3),
    )
preprocess_input = keras.applications.inception_v3.preprocess_input
inputs = keras.Input((224,224, 3))
preprocessed = preprocess_input(inputs)
outputs = inception_model(preprocessed)
feature_extractor = keras.Model(inputs, outputs, name="feature_
extractor")
```

Finally, we will write another function that parses the entire dataset, generates a sequence of frames, extracts features from each frame, and generates a tuple of the feature sequence and the label (real or fake) for the video:

```
def prepare_data(df, data_dir):
    MAX_SEQ_LENGTH = 20
    NUM_FEATURES = 2048
```

```
    num_samples = len(df)
    video_paths = list(df.index)

  # Binarize labels as 0/1
    labels = df["label"].values
    labels = np.array(labels=='FAKE').astype(np.int)


   # Placeholder arrays for frame masks and features
    frame_masks = np.zeros(shape=(num_samples, MAX_SEQ_LENGTH),
dtype="bool")
    frame_features = np.zeros(shape=(num_samples, MAX_SEQ_LENGTH, NUM_
FEATURES), dtype="float32")


  # Parse through each video file
    for idx, path in enumerate(video_paths):
        # Gather all its frames and add a batch dimension.
        frames = parse_video_into_frames(os.path.join(data_dir, path))
        frames = frames[None, ...]


        # Initialize placeholders to store the masks and features of
the current video.
        temp_frame_mask = np.zeros(shape=(1, MAX_SEQ_LENGTH,),
dtype="bool")
        temp_frame_features = np.zeros(shape=(1, MAX_SEQ_LENGTH, NUM_
FEATURES), dtype="float32")


        # Extract features from the frames of the current video.
        for i, batch in enumerate(frames):
            video_length = batch.shape[0]
            length = min(MAX_SEQ_LENGTH, video_length)
            for j in range(length):
                temp_frame_features[i, j, :] = feature_extractor.
predict(batch[None, j, :])
            temp_frame_mask[i, :length] = 1  # 1 = not masked, 0 =
masked

        frame_features[idx,] = temp_frame_features.squeeze()
        frame_masks[idx,] = temp_frame_mask.squeeze()


    return (frame_features, frame_masks), labels
```

To train the model, we need to split the available data into training and test sets (note that the test set provided does not come with labels, so we will not be able to evaluate our model on it). We will use our preprocessing function to prepare the data and convert it into the required form:

```
from sklearn.model_selection import train_test_split
train_set, test_set = train_test_split(train_sample_metadata,test_
size=0.1,random_state=42,stratify=train_sample_metadata['label'])
train_data, train_labels = prepare_data(train_set, "train")
test_data, test_labels = prepare_data(test_set, "test")
```

Now, we can define a model and use this data to train it. Here, we will use the pre-trained model only as a feature extractor. The extracted features will be passed to a downstream model:

```
MAX_SEQ_LENGTH = 20
NUM_FEATURES = 2048
DROPOUT = 0.2
frame_features_input = keras.Input((MAX_SEQ_LENGTH, NUM_FEATURES))
mask_input = keras.Input((MAX_SEQ_LENGTH,), dtype="bool")


x = keras.layers.GRU(32, return_sequences=True)(frame_features_input,
mask=mask_input)
x = keras.layers.GRU(16)(x)
x = keras.layers.GRU(8)(x)
x = keras.layers.Dropout(DROPOUT)(x)
x = keras.layers.Dense(8, activation="relu")(x)
x = keras.layers.Dense(8, activation="relu")(x)
output = keras.layers.Dense(1, activation="sigmoid")(x)

model = keras.Model([frame_features_input, mask_input], output)

model.compile(loss="binary_crossentropy", optimizer="adam",
metrics=["accuracy"])
```

Here, we have used **Gated Recurrent Unit (GRU)** layers followed by two fully connected layers, each with 8 neurons. The last layer contains a single neuron with a sigmoid activation. You can examine the architecture by printing it out:

```
model.summary()
```

Note that, here, we don't use softmax but sigmoid instead. Why do we do this? What we have here is a binary classification problem; the sigmoid will give us the probability that the sequence of frames is a deepfake. If we had a multi-class classification, we would have used softmax instead. Remember that softmax is just sigmoid generalized to multiple classes.

Finally, we train the model. As we use Keras, training is as simple as calling the `fit()` function on the model and passing it the required data:

```
EPOCHS = 10
model.fit(
        [train_data[0], train_data[1]],
        train_labels,
        validation_data=([test_data[0], test_data[1]],test_labels),
        epochs=EPOCHS,
        batch_size=8
    )
```

You should be able to observe the training of the model through each epoch. At every epoch, you can see the loss of the model both over the training data and validation data.

After we have the trained model, we can run the test examples through it and obtain the model prediction. We will obtain the output probability, and if it is greater than our threshold (generally, 0.5), we classify it as fake (label = 1):

```
predicted_labels = []
THRESHOLD = 0.5
for idx in range(len(test_data[0])):
  frame_features = test_data[0][idx]
  frame_mask = test_data[1][idx]
  output_prob = model.predict([frame_features, frame_mask])[0]
  if (output_prob >= THRESHOLD):
    predicted_labels.append(1)
  else:
    predicted_labels.append(0)
```

To examine the performance of our model, we can compare the actual and predicted labels by generating a confusion matrix.

Playing with the model

Recall that in the last section on detecting deepfake images, we included pointers to playing around with the model and tracking performance. We will do the same here for deepfake video classification as well. While such a task may seem redundant, given that we already have a model, in the real world and industry, model tuning is an essential task that data scientists have to handle.

As a learning exercise, you should experiment with the model and determine what the best set of parameters you can have is. Here are some of the things you can experiment with:

- **Image processing**: Note that we parsed videos into frames and resized each image to a 224 x 224 shape. What happens if this shape is changed? How do extremely large sizes (1,048 x 1,048) or extremely small sizes (10 x 10) affect the model? Is the performance affected?

- **Data parameters**: In the preceding walkthrough, while preprocessing the data, we set two important parameters. The first is the *maximum sequence length* that determines how many frames we consider. The second is the *number of features* that controls the length of the feature vector extracted from the pre-trained model. How is the model affected if we change these parameters – both individually and in conjunction with one another?

- **Feature extraction**: We used the paradigms of transfer learning and a pre-trained model to process our input and give us our features. InceptionV3 is one model; however, there are several others that can be used. Do any other models result in better features (as evidenced by better model performance)?

- **Model architecture**: The number of GRU layers, fully connected layers, and the neurons in each layer were chosen arbitrarily. What happens if, instead of 32-16-8 GRU layers, we choose 16-8-8? How about 8-8-8? How about 5 layers, each with 16 neurons (i.e., 16-16-16-16-16)? Similar experiments can be done with the fully connected layers as well. The possibilities here are endless.

- **Training**: We trained our model over 10 epochs. What happens if we train for only 1 epoch? How about 50? 100? Intuitively, you would expect that more epochs leads to better performance, but is that really the case? Additionally, our prediction threshold is set to 0.5. How are the precision and recall affected if we vary this threshold?

This brings us to the end of our discussion on deepfake videos.

Summary

In this chapter, we studied deepfakes, which are synthetic media (images and videos) that are created using deep neural networks. These media often show people in positions that they have not been in and can be used for several nefarious purposes, including misinformation, fraud, and pornography. The impact can be catastrophic; deepfakes can cause political crises and wars, cause widespread panic among the public, facilitate identity theft, and cause defamation and loss of life. After understanding how deepfakes are created, we focused on detecting them. First, we used CNNs to detect deepfake images. Then, we developed a model that parsed deepfake videos into frames and used transfer learning to convert them into vectors, the sequence of which was used for fake or real classification.

Deepfakes are a growing challenge and have tremendous potential for cybercrime. There is a strong demand in the industry for professionals who understand deepfakes, their generation, the social impact they can have, and most importantly, methods to counter them. This chapter provided you with a deep understanding of deepfakes and equips you with the tools and technology needed to detect them.

In the next chapter, we will look at the text counterpart of deepfake images – machine-generated text.

Detecting Machine-Generated Text

In the previous chapter, we discussed deepfakes, which are synthetic media that can depict a person in a video and show the person to be saying or doing things that they did not say or do. Using powerful deep learning methods, it has been possible to create realistic deepfakes that cannot be distinguished from real media. Similar to such deepfakes, machine learning models have also succeeded in creating fake text – text that is generated by a model but appears to be written by a human. While the technology has been used to power chatbots and develop question-answering systems, it has also found its use in several nefarious applications.

Generative text models can be used to enhance bots and fake profiles on social networking sites. Given a prompt text, the model can be used to write messages, posts, and articles, thus adding credibility to the bot. A bot can now pretend to be a real person, and a victim might be fooled because of the realistic-appearing chat messages. These models allow customization by style, tone, sentiment, domain, and even political leaning. It is easily possible to provide a prompt and generate a news-style article; such articles can be used to spread misinformation. Models can be automated and deployed at scale on the internet, which means that there can be millions of fake profiles pretending to be real people, and millions of Twitter accounts generating and posting misleading articles. Detecting automated text is an important problem on the internet today.

This chapter will explore the fundamentals of generative models, how they can be used to create text, and techniques to detect them.

In this chapter, we will cover the following main topics:

- Text generation models
- Naïve detection
- Transformer methods for detecting automated text

By the end of this chapter, you will have a firm understanding of text generation models and approaches to detecting bot-generated text.

Technical requirements

You can find the code files for this chapter on GitHub at `https://github.com/PacktPublishing/10-Machine-Learning-Blueprints-You-Should-Know-for-Cybersecurity/tree/main/Chapter%206`.

Text generation models

In the previous chapter, we saw how machine learning models can be trained to generate images of people. The images generated were so realistic that it was impossible in most cases to tell them apart from real images with the naked eye. Along similar lines, machine learning models have made great progress in the area of text generation as well. It is now possible to generate high-quality text in an automated fashion using deep learning models. Just like images, this text is so well written that it is not possible to distinguish it from human-generated text.

Fundamentally, a language model is a machine learning system that is able to look at a part of a sentence and predict what comes next. The words predicted are appended to the existing sentence, and this newly formed sentence is used to predict what will come next. The process continues recursively until a specific token denoting the end of the text is generated. Note that when we say that the next word is predicted, in reality, the model generates a probability distribution over possible output words. Language models can also operate at the character level.

Most text generation models take in a prompt text as input. Trained on massive datasets (such as all Wikipedia articles or entire books), the models have learned to produce text based on these prompts. Training on different kinds of text (stories, biographies, technical articles, and news articles) enables models to generate those specific kinds of text.

To see the power of AI-based text generation with your own eyes, explore the open source text generator called **Grover**. This is a tool that was produced by researchers at the University of Washington and allows you to produce a real-looking news article based on any given prompt. The website provides an interface as shown in the following figure, where you can provide a prompt title and an author name:

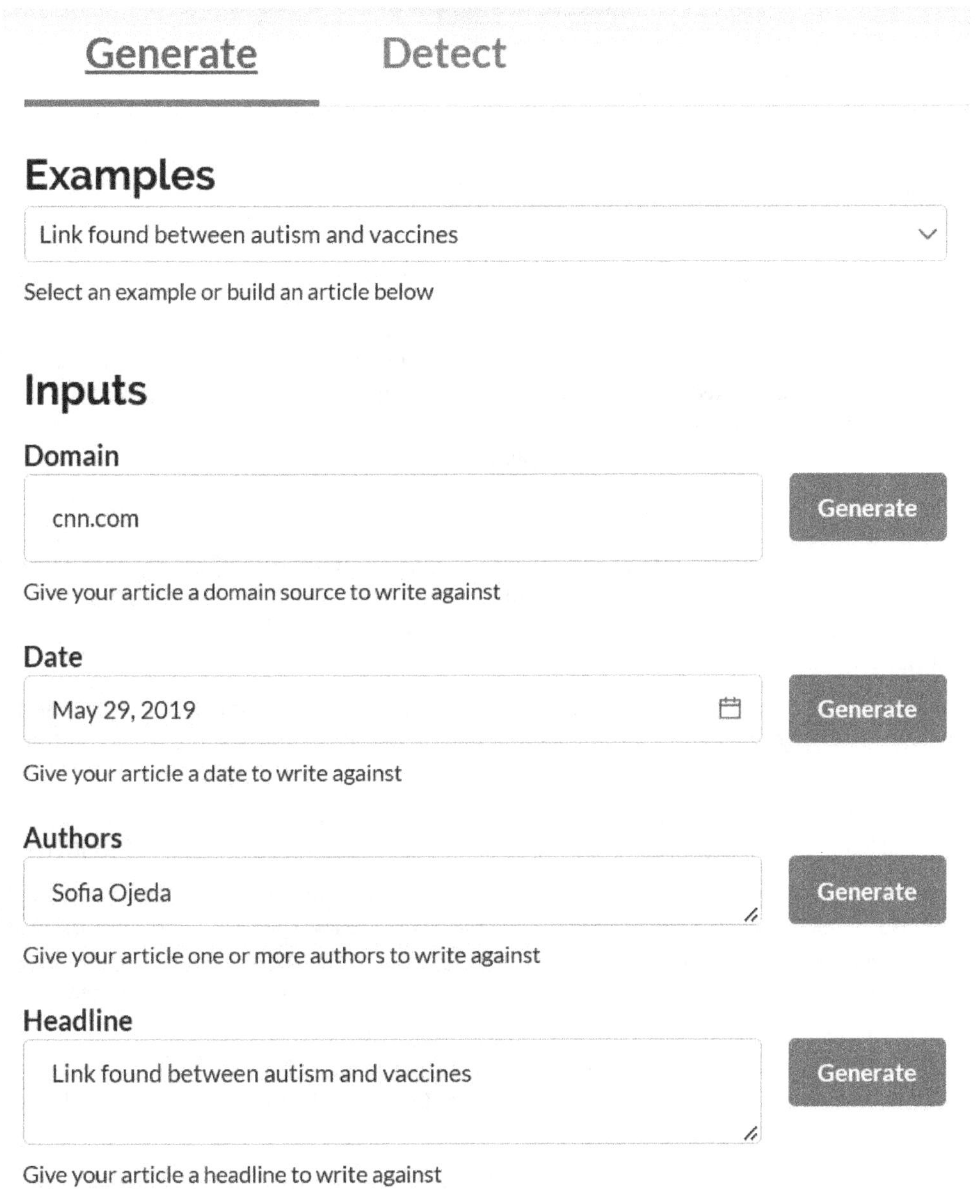

Generate   Detect

Examples

Link found between autism and vaccines   ⌄

Select an example or build an article below

Inputs

Domain

cnn.com   **Generate**

Give your article a domain source to write against

Date

May 29, 2019   🗓   **Generate**

Give your article a date to write against

Authors

Sofia Ojeda   **Generate**

Give your article one or more authors to write against

Headline

Link found between autism and vaccines   **Generate**

Give your article a headline to write against

Figure 6.1 – The Grover interface to generate text

Once you set these parameters, Grover will generate an article that appears real. Here is the article
that was generated with our given prompt:

Link found between autism and vaccines

May 29, 2019 - Sofia Ojeda

Written by Sofia Ojeda, CNN

A paper published in the journal Genetics and Human Behavior claims to find a "significant link" between autism and a type of booster vaccine given in childhood. The paper focuses on the mercury-based preservative thimerosal and considers the possibility that the injectable vaccine contributes to autism-related symptoms. These include poor social skills, language delays and seizures, according to the paper.

Thimerosal, which is commonly used as a preservative in vaccines, was phased out of children's vaccines in 2010. In 2010, the Food and Drug Administration declared that thimerosal was safe. "We don't know of any adverse effects associated with the endocrine-disrupting, mercury-containing preservative in vaccines, and believe that this outcome is beneficial to public health," said Dr. Lisa M. Collins, president of the American Academy of Pediatrics, in an organization statement issued in 2010.

The new study, led by researchers at Tel Aviv University, is based on case studies of two women with autistic children and one non-autistic mother. The researchers found that autism symptoms tended to worsen after only one dose of the thimerosal-containing hepatitis B booster vaccine. "The use of unadjuvanted Hepatitis B booster vaccines, against which the only adverse effects are fatigue, headache, cough, and fever, remains a substantial and unrefined source of vaccine-related injury and disability worldwide," the study authors wrote.

Using a mathematical model, the study authors described multiple risks associated with the use of unadjuvanted injectable vaccines. Among them, vaccines increased the risk of seizures, such as those associated with autism; autoimmune disorders and cancer.

"There are other types of vaccines that pose similar risk to autism, including rotavirus vaccines and the polio vaccine," said lead author Professor Uri Ayalon, head of the pediatrics department at Tel Aviv University.

The authors also found that the autism risk increased by twofold in children aged 3 to 10 who received the hepatitis B booster vaccine against Hepatitis B, regardless of the following vaccination schedules: aged 9 months or older with only one dose; or under 4 months old and only one dose. "Since these vaccines are widely used, regardless of age, it may be important to limit its utilization," the researchers wrote.

Lead author Dr. Yonatan Schulmann said there were no apparent risks associated with a standard influenza vaccination. "The flu vaccine probably represents an acceptable source of vaccine-related injury and disability," he said. "This is not true for most vaccines. The flu vaccine is relatively inexpensive (free of charges) and has no significant health effects," he said.

The timing of vaccination is also important, said Schulmann. "Autism spectrum disorders are most often diagnosed in early adolescence, the upper age range at which it is most likely that vaccination data is available," he said. Furthermore, the authors said they found no clear differences between children who received hepatitis B vaccine against Hepatitis B and other children.

Note how the article has the stylistic features that you would typically expect from journalistic writings. The sentence construction is grammatically correct and the whole text reads as a coherent article. There are quotes from researchers and professors who are subject matter experts, complete with statistics and experimental results cited. Overall, the article could pass off as something written by a human.

Understanding GPT

GPT stands for **Generative Pretrained Transformer**, and GPT models have dazzled the NLP world because they can generate coherent essays that are beyond those produced by traditional language models such as those based on **Recurrent Neural Networks (RNNs)**. GPT models are also based on the transformer architecture (recall the BERT architecture that we used for malware detection was also based on the transformer).

Recall the concepts of attention that we introduced in *Chapter 3, Malware Detection Using Transformers and BERT*. We introduced two kinds of blocks – the encoder and decoder – both of which were built using transformers that leveraged the attention mechanism. The transformer encoder had a self-attention layer followed by a fully connected feed-forward neural network. The decoder layer was similar except that it had an additional masked self-attention layer that ensured that the transformer did not attend to the future tokens (which would defeat the purpose of the language model). For example, if the decoder decodes the fourth word, it will attend to all words up to the third predicted word and all the words in the input.

In general, GPT models use only the decoder blocks, which are stacked one after the other. When a token is fed into the model, it is converted into an embedding representation using a matrix lookup. Additionally, a positional encoding is added to it to indicate the sequence of words/tokens. The two matrices (embedding and positional encoding) are parts of the pretrained models we use. When the first token is passed to the model, it gets converted into a vector using the embedding lookup and positional encoding matrices. It passes through the first decoder block, which performs self-attention, passes the output to the neural network layer, and forwards the output to the next decoder block.

After processing by the final decoder, the output vector is multiplied with the embedding matrix to obtain a probability distribution over the output token to be produced. This probability distribution can be used to select the next word. The most straightforward strategy is to choose the word with the highest probability – however, we run the risk of being stuck in a loop. For instance, if the tokens produced so far are "*The man and*" and we always select the word with the highest probability, we might end up producing "*The man and the man and the man and the man.....*" indefinitely.

To avoid this, we apply a top-K sampling. We select the top K words (based on the probability) and sample a word from them, where words with a higher score have a higher chance of being selected. Since this process is non-deterministic, the model does not end up in the loop of choosing the same set of words again and again. The process continues until a certain number of tokens has been produced, or the end-of-string token is found.

Generation by GPT models can be either conditional or unconditional. To see generation in action, we can use the Write with Transformer (`https://transformer.huggingface.co/doc/gpt2-large`) web app developed by Hugging Face, which uses GPT-2. The website allows you to simulate both conditional and unconditional generation.

In conditional generation, we provide the model with a set of words as a prompt, which is used to seed the generation. This initial set of words provides the context used to drive the rest of the text, as shown:

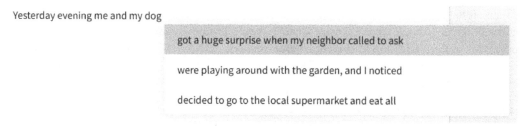

Figure 6.2 – Generating text with a prompt

On the other hand, in unconditional generation, we just provide the `<s>` token, which is used to indicate the start of a string, and allow the model to freely produce what it wants. If you press the *Tab* key on Write With Transformer, you should see such unconditional samples generated:

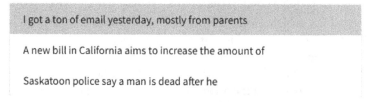

Figure 6.3 – Generating text without prompts

There have been multiple versions of GPT models released by OpenAI, the latest one that has made the news being ChatGPT, based on GPT 3.5. In an upcoming section, we will use ChatGPT to create our own dataset of fake news.

Naïve detection

In this section, we will focus on naïve methods for detecting bot-generated text. We will first create our own dataset, extract features, and then apply machine learning models to determine whether a particular text is machine-generated or not.

Creating the dataset

The task we will focus on is detecting bot-generated fake news. However, the concepts and techniques we will learn are fairly generic and can be applied to parallel tasks such as detecting bot-generated tweets, reviews, posts, and so on. As such a dataset is not readily available to the public, we will create our own.

How are we creating our dataset? We will use the News Aggregator dataset (`https://archive.ics.uci.edu/ml/datasets/News+Aggregator`) from the UCI Dataset Repository. The dataset contains a set of news articles (that is, links to the articles on the web). We will scrape these articles, and these are our human-generated articles. Then, we will use the article title as a prompt to seed generation by GPT-2, and generate an article that will be on the same theme and topic, but generated by GPT-2! This makes up our positive class.

Scraping real articles

The News Aggregator dataset from UCI contains information on over 420k news articles. It was developed for research purposes by scientists at the Roma Tre University in Italy. News articles span multiple categories such as business, health, entertainment, and science and technology. For each article, we have the title and the URL of the article online. You will need to download the dataset from the UCI Machine Learning Repository website (`https://archive.ics.uci.edu/ml/datasets/News+Aggregator`).

Take a look at the data using the `head()` functionality (note that you will have to change the path according to how you store the file locally):

```
import pandas as pd

path = "/content/UCI-News-Aggregator-Classifier/data/uci-news-
aggregator.csv"
df = pd.read_csv(path)
df.head()
```

This will show you the first five rows of the DataFrame. As you can see in the following screenshot, we have an ID to refer to each row and the title and URL of the news article. We also have the hostname (the website where the article appeared) and the timestamp, which denotes the time when the news was published. The **STORY** field contains an ID that is used to indicate a cluster containing similar news stories.

	ID	TITLE	URL	PUBLISHER	CATEGORY		STORY	HOSTNAME	TIMESTAMP
0	1	Fed official says weak data caused by weather...	http://www.latimes.com/business/money/la-fi-mo...	Los Angeles Times	b	ddUyU0VZz0BRneMioxUPQVP6sIxvM	www.latimes.com	1394470370698
1	2	Fed's Charles Plosser sees high bar for change...	http://www.livemint.com/Politics/H2EvwJSK2VE6O...	Livemint	b	ddUyU0VZz0BRneMioxUPQVP6sIxvM	www.livemint.com	1394470371207
2	3	US open: Stocks fall after Fed official hints ...	http://www.ifamagazine.com/news/us-open-stocks...	IFA Magazine	b	ddUyU0VZz0BRneMioxUPQVP6sIxvM	www.ifamagazine.com	1394470371550
3	4	Fed risks falling 'behind the curve', Charles ...	http://www.ifamagazine.com/news/fed-risks-fall...	IFA Magazine	b	ddUyU0VZz0BRneMioxUPQVP6sIxvM	www.ifamagazine.com	1394470371793
4	5	Fed's Plosser: Nasty Weather Has Curbed Job Gr...	http://www.moneynews.com/Economy/federal-reser...	Moneynews	b	ddUyU0VZz0BRneMioxUPQVP6sIxvM	www.moneynews.com	1394470372027

Figure 6.4 – UCI News Aggregator data

Let us take a look at the distribution of the articles across categories:

```
df["CATEGORY"].value_counts().plot(kind = 'bar')
```

This will produce the following result:

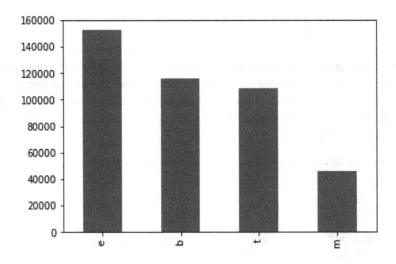

Figure 6.5 – News article distribution by category

From the documentation, we see that the categories **e**, **b**, **t**, and **m** represent entertainment, business, technology, and health, respectively. Entertainment has the highest number of articles, followed by business and technology (which are similar), and health has the least.

Similarly, we can also inspect the top domains where the articles come from:

```
df["HOSTNAME"].value_counts()[:20].plot(kind = 'bar')
```

You will get the following output:

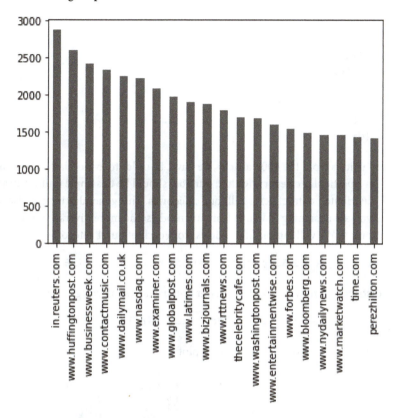

Figure 6.6 – Distribution of news articles across sources

In order to scrape the article from the website, we would need to simulate a browser session using a browser tool such as Selenium, find the article text by parsing the HTML source, and then extract it. Fortunately, there is a library in Python that does all of this for us. The Newspaper Python package (https://github.com/codelucas/newspaper/) provides an interface for downloading and parsing news articles. It can extract text, keywords, author names, summaries, and images from the HTML source of an article. It has support for multiple languages including English, Spanish, Russian, and German. You can also use a general-purpose web scraping library such as BeautifulSoup, but the Newspaper library is designed specifically to capture news articles and hence provides a lot of functions that we would have had to write custom if using BeautifulSoup.

To create our dataset of real articles, we will iterate through the News Aggregator DataFrame and use the Newspaper library to extract the text for each article. Note that the dataset has upward of 420k articles – for the purposes of demonstration, we will sample 1,000 articles randomly from the dataset. For each article, we will use the Newspaper library to scrape the text. We will create a directory to hold these articles.

First, let us create the directory structure:

```python
import os
root = "./articles"
fake = os.path.join(root, "fake")
real = os.path.join(root, "real")

for dir in [root, real, fake]:
  if not os.path.exists(dir):
    os.mkdir(dir)
```

Now, let us sample articles from the 400k articles we have. In order to avoid bias and overfitting, we should not focus on a particular category. Rather, our goal should be to sample uniformly at random so we have a well-distributed dataset across all four categories. This general principle also applies to other areas where you are designing machine learning models; the more diverse your dataset is, the better the generalization. We will sample 250 articles from each of our 4 categories:

```python
df2 = df.groupby('CATEGORY').apply(lambda x: x.sample(250))
```

If you check the distribution now, you will see that it is equal across all categories:

```python
df2["CATEGORY"].value_counts().plot(kind='bar')
```

You can see the distribution clearly in the following plot:

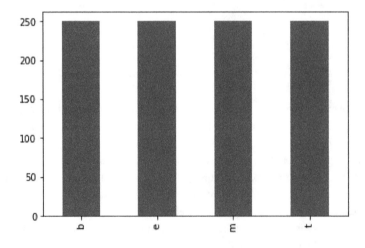

Figure 6.7 – Distribution of sampled articles

We will now iterate through this DataFrame and scrape each article. We will scrape the article, read the text, and save it into a file in the real directory we created earlier. Note that this is essentially a web scraper – as different websites have different HTML structures, the newspaper library may hit some errors. Certain websites may also block scrapers. For such articles, we will print out a message with the article URL. In practice, when such a situation is encountered, data scientists will fill the gap manually if the number of missing articles is small enough:

```
from newspaper import Article

URL_LIST = df2["URL"].tolist()
TITLE_LIST = df2["TITLE"].tolist()
for id_url, article_url in enumerate(URL_LIST):
  article = Article(article_url)
  try:
    # Download and parse article
    article.download()
    article.parse()
    text = article.text

    # Save to file
    filename = os.path.join(real, "Article_{}.txt".format(id_url))
    article_title = TITLE_LIST[id_url]
    with open(filename, "w") as text_file:
      text_file.write(" %s \n %s" % (article_title, text))

  except:
    print("Could not download the article at: {}".format(article_url))
```

Now we have our real articles downloaded locally. It's time to get into the good stuff – creating our set of fake articles!

Using GPT to create a dataset

In this section, we will use GPT-3 to create our own dataset of machine-generated text. OpenAI, a San Francisco-based artificial intelligence research lab, developed GPT-3, a pretrained universal language model that utilizes deep learning transformers to create text that is remarkably human-like. Released in 2020, GPT-3 has made headlines in various industries, as its potential use cases are virtually limitless. With the help of the GPT-3 API family and ChatGPT, individuals have used it to write fiction and poetry, code websites, respond to customer feedback, improve grammar, translate languages, generate dialog, optimize tax deductions, and automate A/B testing, among other things. The model's high-quality results have impressed many.

We can use the `transformers` library from HuggingFace to download and run inference on ChatGPT models. To do this, we can first load the model as follows:

```
from transformers import pipeline
generator = pipeline('text-generation',
                     model='EleutherAI/gpt-neo-2.7B')
```

This will download the model for you locally. Note that this involves downloading a sizeable model from the online repository, and hence will take quite some time. The time taken to execute will depend on your system usage, resources, and network speed at the time.

We can generate a sample text using this new model. For example, if we want to generate a poem about flowers, we can do the following:

```
chat_prompt = 'Generate a five-line poem about flowers'
model_output = generator(prompt,
                    max_length=100)
response = model_output[0]['generated_text']
print(response)
```

And this gave me the following poem (note that the results may differ for you):

```
Flowers bloom in gardens bright,
Their petals open to the light,
Their fragrance sweet and pure,
A colorful feast for eyes to lure,
Nature's art, forever to endure.
```

We have already downloaded and initialized the model we want. Now, we can iterate through our list of article titles and generate articles one by one by passing the title as a seed prefix. Just like the scraped articles, each article must be saved into a text file so that we can later access it for training:

```
for id_title, title in enumerate(TITLE_LIST):
    # Generate the article
    article = generator(title, max_length = 500)[0]["generated_text"]

    # Save to file
    filename = os.path.join(fake, "Article_{}.txt".format(id_url))
    with open(filename, "w") as text_file:
        text_file.write(" %s \n %s" % (title, text))
```

All that is left now is to read all of the data we have into a common array or list, which can then be used in all of our experiments. We will read each file in the real directory and add it to an array. At the same time, we will keep appending 0 (indicating a real article) to another array that holds labels. We will repeat the same process with the fake articles and append 1 as the label:

```
X = []
Y = []

for file in os.listdir(real):
    try:
        with open(file, "r") as article_file:
```

```
        article = file.read()
        X.append(article)
        Y.append(0)
    except:
      print("Error reading: {}".format(file))
      continue


for file in os.listdir(fake):
  try:
    with open(file, "r") as article_file:
      article = file.read()
      X.append(article)
      Y.append(1)
  except:
    print("Error reading: {}".format(file))
    continue
```

Now, we have our text in the X list and associated labels in the Y list. Our dataset is ready!

Feature exploration

Now that we have our dataset, we want to build a machine learning model to detect bot-generated news articles. Recall that machine learning algorithms are mathematical models and, therefore, operate on numbers; they cannot operate directly on text! Let us now extract some features from the text.

This section will focus on hand-crafting features – the process where subject matter experts theorize potential differences between the two classes and build features that will effectively capture the differences. There is no unified technique for doing this; data scientists experiment with several features based on domain knowledge to identify the best ones.

Here, we are concerned with text data – so let us engineer a few features from that domain. Prior work in NLP and linguistics has analyzed human writing and identified certain characteristics. We will engineer three features based on prior research.

Function words

These are supporting words in the text that do not contribute to meaning but add continuity and flow to the sentence. They are generally determiners (*the, an, many, a little,* and *none*), conjunctions (*and* and *but*), prepositions (*around, within,* and *on*), pronouns (*he, her,* and *their*), auxiliary verbs (*be, have,* and *do*), modal auxiliary (*can, should, could,* and *would*), qualifiers (*really* and *quite*), or question words (*how* and *why*). Linguistic studies have shown that every human uses these unpredictably, so there might be randomness in the usage pattern. As our feature, we will count the number of function words that we see in the sentence, and then normalize it by the length of the sentence in words.

We will use a file that contains a list of the top function words and read the list of all function words. Then, we will count the function words in each text and normalize this count by the length. We will wrap this up in a function that can be used to featurize multiple instances of text. Note that as the list of function words would be the same for all texts, we do not need to repeat it in each function call – we will keep that part outside the function:

```python
FUNCTION_WORD_FILE = '../static/function_words.txt'
with open(FUNCTION_WORD_FILE,'r') as fwf:
  k = fwf.readlines()
  func_words = [w.rstrip() for w in k]

  #There might be duplicates!
  func_words = list(set(func_words))

def calculate_function_words(text):
  function_word_counter = 0
  text_length = len(text.split(' '))
  for word in func_words:
    function_word_counter = function_word_counter + text.count(word)

  if text_length == 0:
    feature = 0
  else:
    feature = function_word_counter / total_length

  return feature
```

Punctuation

Punctuation symbols (commas, periods, question marks, exclamations, and semi-colons) set the tone of the text and inform how it should be read. Prior research has shown that the count of punctuation symbols may be an important feature in detecting bot-generated text. We will first compile a list of punctuation symbols (readily available in the Python `string` package). Similar to the function words, we will count the occurrences of punctuation and normalize them by length. Note that this time, however, we need to normalize by the length in terms of the number of characters as opposed to words:

```python
def calculate_punctuation(text):
  punctuations = =[ k for k in string.punctuation]
  punctuation_counter = 0
  total_length = len(text.split())
```

```
for punc in punctuations:
    punctuation_counter = punctuation_counter + text.count(punc)

if text_length == 0:
    feature = 0
else:
    feature = punctuation_counter / total_length

return feature
```

Readability

Research in early childhood education has studied text in detail and derived several metrics that indicate how readable a particular blob of text is. These metrics analyze the vocabulary and complexity of the text and determine the ease with which a reader can read and understand the text. There are several measures of readability defined in prior literature (https://en.wikipedia.org/wiki/Readability), but we will be using the most popular one called the **Automated Readability Index** (**ARI**) (https://readabilityformulas.com/automated-readability-index.php). It depends on two factors – word difficulty (the number of letters per word) and sentence difficulty (the number of words per sentence), and is calculated as follows:

$$ARI = 4.71 \left(\frac{\text{\# characters}}{\text{\# words}} \right) + 0.5 \left(\frac{\text{\# words}}{\text{\# sentences}} \right) - 21.43$$

In theory, the ARI represents the approximate age needed to understand the text. We will now develop a function that calculates the ARI for our input text, and wrap it into a function like we did for the previous features:

```
def calculate_ari(text):
    chars = len(text.split())
    words = len(text.split(' '))
    sentences = len(text.split('.'))

    if words == 0 or sentences == 0:
        feature = 0
    else:
        feature = 4.71* (chars / words) + 0.5* (words / sentences) - 21.43
    return feature
```

This completes our discussion of naive feature extraction. In the next section, we will use these features to train and evaluate machine learning models.

Using machine learning models for detecting text

We have now hand-crafted three different features: punctuation counts, function word counts, and the readability index. We also defined functions for each. Now, we are ready to apply these to our dataset and build models. Recall that the X array contains all of our text. We want to represent each text sample using a three-element vector (as we have three features):

```
X_Features = []
for x in X:
  feature_vector = []
  feature_vector.append(calculate_function_words(x))
  feature_vector.append(calculate_punctuation(x))
  feature_vector.append(calculate_ari(x))
  X_Features.append(feature_vector)
```

Now, each text sample is represented by a three-element vector in X_Features. The first, second, and third elements represent the normalized function word count, punctuation count, and ARI, respectively. Note that this order is arbitrary – you may choose your own order as it does not affect the final model.

Our features are ready, so now we will do the usual. We begin by splitting our data into training and test sets. We then fit a model on the training data and evaluate its performance on the test data. In previous chapters, we used the confusion matrix function to plot the confusion matrix and visually observe the true positives, false positives, true negatives, and false negatives. We will now build another function on top of it that will take in these values and calculate metrics of interest. We will calculate the true positives, false positives, true negatives, and false negatives, and then calculate the accuracy, precision, recall, and F1 score. We will return all of these as a dictionary:

```
from sklearn.metrics import confusion_matrix

def evaluate_model(actual, predicted):
  confusion = confusion_matrix(actual, predicted)
  tn, fp, fn, tp = confusion.ravel()

  total = tp + fp + tn + fn

  accuracy = 100 * (tp + tn) / total
  if tp + fp != 0:
    precision = tp / (tp + fp)
  else:
    precision = 0

  if tp + fn != 0:
    recall = tp / (tp + fn)
```

```
    else:
      recall = 0

    if precision == 0 or recall == 0:
      f1 = 0
    else:
      f1 = 2 * precision * recall / (precision + recall)

    evaluation = { 'accuracy': accuracy,'precision': precision,'recall':
recall,'f1': f1}

    return evaluation
```

Let us split the data into training and testing:

```
from sklearn.model_selection import train_test_split
X_train, X_test, Y_train, Y_test = train_test_split(X_Features, Y)
```

Now, we will fit a model on the training data, and evaluate its performance on the test data. Here, we will use random forests, logistic regression, SVM, and a deep neural network.

The logistic regression classifier is a statistical model that expresses the probability of an input belonging to a particular class as a linear combination of features. Specifically, the model produces a linear combination of inputs (just like linear regression) and applies a sigmoid to this combination to obtain an output probability:

```
from sklearn.linear_model import LogisticRegression
model = LogisticRegression()
model.fit(X_train, Y_train)
Y_predicted = model.predict(X_test)
print(evaluate_model(Y_test, Y_pred))
```

Random forests are ensemble classifiers consisting of multiple decision trees. Each tree is a hierarchical structure with nodes as conditions and leaves as class labels. A classification label is derived by following the path of the tree through the root. The random forest contains multiple such trees, each trained on a random sample of data and features:

```
from sklearn.ensemble import RandomForestClassifier
model = RandomForestClassifier(n_estimators = 100)
model.fit(X_train, Y_train)
Y_predicted = model.predict(X_test)
print(evaluate_model(Y_test, Y_pred))
```

A **multilayer perceptron** (**MLP**) is a fully connected deep neural network, with multiple hidden layers. The input data undergoes transformations through these layers, and the final layer is a sigmoid or softmax function, which generates the probability of the data belonging to a particular class:

```
from sklearn.neural_network import MLPClassifier
model = MLPClassifier(hidden_layer_sizes = (50, 25, 10),max_iter =
100,activation = 'relu',solver = 'adam',random_state = 123)
model.fit(X_train, Y_train)
Y_predicted = model.predict(X_test)
print(evaluate_model(Y_test, Y_pred))
```

The SVM constructs a decision boundary between two classes such that the best classification accuracy is obtained. In case the boundary is not linear, the SVM transforms the features into a higher dimensional space and obtains a non-linear boundary:

```
from sklearn import svm
model = svm.SVC(kernel='linear')
model.fit(X_train, Y_train)
Y_predicted = model.predict(X_test)
print(evaluate_model(Y_test, Y_pred))
```

Running this code should print out the evaluation dictionaries for each model, which tells you the accuracy, recall, and precision. You can also plot the confusion matrix (as we did in previous chapters) to visually see the false positives and negatives, and get an overall sense of how good the model is.

Playing around with the model

We have explored here only three features – however, the possibilities for hand-crafted features are endless. I encourage you to experiment by adding more features to the mix. Examples of some features are as follows:

- Length of the text

- Number of proper nouns

- Number of numeric characters

- Average sentence length

- Number of times the letter q was used

This is certainly not an exhaustive list, and you should experiment by adding other features to see whether the model's performance improves.

Automatic feature extraction

In the previous section, we discussed how features can be engineered from text. However, hand-crafting features might not always be the best idea. This is because it requires expert knowledge. In this case, data scientists or machine learning engineers alone will not be able to design these features – they will need experts from linguistics and language studies to identify the nuances of language and suggest appropriate features such as the readability index. Additionally, the process is time-consuming; each feature has to be identified, implemented, and tested one after the other.

We will now explore some methods for automatic feature extraction from text. This means that we do not manually design features such as the punctuation count, readability index, and so on. We will use existing models and techniques, which can take in the input text and generate a feature vector for us.

TF-IDF

Term Frequency – Inverse Document Frequency (TF-IDF) is a commonly used technique in natural language processing to convert text into numeric features. Every word in the text is assigned a score that indicates how important the word is in that text. This is done by multiplying two metrics:

- **Term Frequency**: How frequently does the word appear in the text sample? This can be normalized by the length of the text in words, as texts that differ in length by a large number can cause skews. The term frequency measures how common a word is in this particular text.

- **Inverse Document Frequency**: How frequently does the word appear in the rest of the corpus? First, the number of text samples containing this word is obtained. The total number of samples is divided by this number. Simply put, IDF is the inverse of the fraction of text samples containing the word. IDF measures how common the word is in the rest of the corpus.

For every word in each text, the TF-IDF score is a statistical measure of the importance of the word to the sentence. A word that is common in a text but rare in the rest of the corpus is surely important and a distinguishing characteristic of the text, and will have a high TF-IDF score. Alternately, a word that is very common in the corpus (that is, present in nearly all text samples) will not be a distinguishing one – it will have a low TF-IDF score.

In order to convert the text into a vector, we first calculate the TF-IDF score of each word in each text. Then, we replace the word with a sequence of TF-IDF scores corresponding to the words. The `scikit-learn` library provides us with an implementation of TF-IDF vectorization out of the box.

Note a fine nuance here: the goal of our experiment is to build a model for bot detection that can be used to classify new text as being generated by bots or not. Thus, when we are training, we have no idea about the test data that will come in the future. To ensure that we simulate this, we will do the TF-IDF score calculation over only the training data. When we vectorize the test data, we will simply use the calculated scores as a lookup:

```
from sklearn.feature_extraction.text import TfidfVectorizer
tf_idf = TfidfVectorizer()
```

```
X_train_TFIDF = tf_idf.fit_transform(X_train)
X_test_TFIDF = tf_idf.transform(X_test)
```

You can manually inspect a few samples from the generated list. What do they look like?

Now that we have the feature vectors, we can use them to train the classification models. The overall procedure remains the same: initialize a model, fit a model on the training data, and evaluate it on the testing data. The MLP example is shown here; however, you could replace this with any of the models we discussed:

```
from sklearn.neural_network import MLPClassifier
model = MLPClassifier(hidden_layer_sizes = (300, 200, 100),max_iter =
100,activation = 'relu',solver = 'adam',random_state = 123)
model.fit(X_train_TFIDF, Y_train)
Y_predicted = model.predict(X_test_TFIDF)
print(evaluate_model(Y_test, Y_pred))
```

How does the performance of this model compare to the performance of the same model with handcrafted features? How about the performance of the other models?

Word embeddings

The TF-IDF approach is considered to be what we call a *bag of words* approach in machine learning terms. Each word is scored based on its presence, irrespective of the order in which it appears. Word embeddings are numeric representations of words assigned such that words that are similar in meaning have similar embeddings – the numeric representations are close to each other in the feature space. The most fundamental technique used to to generate word embeddings is called **Word2Vec**.

Word2Vec embeddings are produced by a shallow neural network. Recall that the last layer of a classification model is a sigmoid or softmax layer for producing an output probability distribution. This softmax layer operates on the features it receives from the pre-final layer – these features can be treated as high-dimensional representations of the input. If we chop off the last layer, the neural network without the classification layer can be used to extract these embeddings.

Word2Vec can work in one of two ways:

- **Continuous Bag of Words**: A neural network model is trained to predict the next word in a sentence. Input sentences are broken down to generate training examples. For example, if the text corpus contains the sentence *I went to walk the dog*, then X = *I went to walk the* and Y = *dog* would be one training example.

- **Skip-Gram**: This is the more widely used technique. Instead of predicting the target word, we train a model to predict the surrounding words. For example, if the text corpus contains the sentence *I went to walk the dog*, then our input would be *walk* and the output would be a prediction (or probabilistic prediction) of the surrounding two or more words. Because of this design, the model learns to generate similar embeddings for similar words. After the model is trained, we can pass the word of interest as an input, and use the features of the final layer as our embedding.

Note that while this is still a classification task, it is not supervised learning. Rather, it is a self-supervised approach. We have no ground truth, but by framing the problem uniquely, we generate our own ground truth.

We will now build our word embedding model using the `gensim` Python library. We will fit the model on our training data, and then vectorize each sentence using the embeddings. After we have the vectors, we can fit and evaluate the models.

First, we fit the model on training data. Because of the way Word2Vec operates, we need to combine our texts into a list of sentences and then tokenize it into words:

```python
import nltk
nltk.download('punkt')
corpus = []
for x in X_train:
    # Split into sentences
    sentences_tokens = nltk.sent_tokenize(x)
    # Split each sentence into words
    word_tokens = [nltk.word_tokenize(sent) for sent in sentences_
tokens]
    # Add to corpus
    corpus = corpus + word_tokens
```

Now, we can fit the embedding model. By passing in the `vector_size` parameter, we control the size of the generated embedding. The larger the size, the more the expressive the power of the embeddings:

```python
from gensim.models import Word2Vec
model = Word2Vec(corpus, min_count=1, vector_size = 30)
```

We now have the embedding model and can start using it to tokenize the text. Here, we have two strategies. One strategy is that we can calculate the embedding for all the words in the text and simply average them to find the mean embedding for the text. Here's how we would do this:

```python
X_train_vector_mean = []
for x in X_train:
    # Create a 30-element vector with all zeroes
    vector = [0 for _ in range(30)]
    # Create a vector for out-of-vocab words
    oov = [0 for _ in range(30)]

    words = x.split(' ')
    for word in words:
        if word in model.wv.vocab:
```

```
        # Word is present in the vocab
        vector = np.sum([vector, model[word]], axis = 0)
    else:
        # Out of Vocabulary
        vector = np.sum([vector, oov], axis = 0)

  # Calculate the mean
  mean_vector = vector / len(words)
  X_train_vector_mean.append(mean_vector)
```

The X_train_vector_mean array now holds an embedding representation for each text in our corpus. The same process can be repeated to generate the feature set with test data.

The second strategy is, instead of averaging the vectors, we append them one after the other. This retains more expressive power as it takes into account the order of words in the sentence. However, each text will have a different length and we require a fixed-size vector. Therefore, we take only a fixed number of words from the text and concatenate their embeddings.

Here, we set the maximum number of words to be 40. If a text has more than 40 words, we will consider only the first 40. If it has less than 40 words, we will consider all of the words and pad the remaining elements of the vector with zeros:

```
X_train_vector_appended = []
max_words = 40
for x in X_train:
  words = x.split(' ')
  num_words = max(max_words, len(words))
  feature_vector = []
  for word in words[:num_words]:
    if word in model.wv.vocab:
      # Word is present in the vocab
      vector = np.sum([vector, model[word]], axis = 0)
    else:
      # Out of Vocabulary
      vector = np.sum([vector, oov], axis = 0)
    feature_vector = feature_vector + vector

  if num_words < max_words:
    pads = [0 for _ in range(30*(max_words-num_words))]
    feature_vector = feature_vector + pads

  X_train_vector_appended.append(feature_vector)
```

The same code snippet can be repeated with the test data as well. Remember that the approach you use (averaging or appending) has to be consistent across training and testing data.

Now that the features are ready, we train and evaluate the model as usual. Here's how you would do it with an MLP; this is easily extensible to other models we have seen:

```
from sklearn.neural_network import MLPClassifier
model = MLPClassifier(hidden_layer_sizes = (1000, 700, 500, 200),max_
iter = 100,activation = 'relu',solver = 'adam',random_state = 123)
model.fit(X_train_vector_appended, Y_train)
Y_predicted = model.predict(X_test_vector_appended)
print(evaluate_model(Y_test, Y_pred))
```

Note that, here, the dimensions of the hidden layers we passed to the model are different from before. In the very first example with hand-crafted features, our feature vector was only three-dimensional. However, in this instance, every text instance will be represented by 40 words, and each word represented by a 30-dimensional embedding, meaning that the feature vector has 1,200 elements. The higher number of neurons in the hidden layer helps handle the high-dimensional feature space.

As an exercise, you are encouraged to experiment with three changes and check whether there is an improvement in the model performance:

- The size of the word embeddings, which has been set to 30 for now. What happens to the model performance as you increase or decrease this number?

- The number of words has been chosen as 0. What happens if this is reduced or increased?

- Using the MLP, how does the model performance change as you vary the number of layers?

- Combine word embeddings and TF-IDF. Instead of a simple average, calculate a weighted average where the embedding for each word is weighted by the TF-IDF score. This will ensure that more important words influence the average more. How does this affect model performance?

Transformer methods for detecting automated text

In the previous sections, we have used traditional hand-crafted features, automated bag of words features, as well as embedding representations for text classification. We saw the power of BERT as a language model in the previous chapter. While describing BERT, we referenced that the embeddings generated by BERT can be used for downstream classification tasks. In this section, we will extract BERT embeddings for our classification task.

The embeddings generated by BERT are different from those generated by the Word2Vec model. Recall that in BERT, we use the masked language model and a transformer-based architecture based on attention. This means that the embedding of a word depends on the context in which it occurs; based on the surrounding words, BERT knows which other words to pay attention to and generate the embedding.

In traditional word embeddings, a word will have the same embedding, irrespective of the context. The word *match* will have the same embedding in the sentence *They were a perfect match!* and *I lit a match last night.* BERT, on the other hand, conditions the embeddings based on context. The word *match* would have different embeddings in these two sentences.

Recall that we have already used BERT once, for malware detection. There are two major differences in how we use it now versus when we implemented it for malware detection:

- Previously, we used BERT in the fine-tuning mode. This means that we used the entire transformer architecture initialized with pretrained weights, added a neural network on top of it, and trained the whole model end to end. The pretrained model enabled learning sequence features, and the fine-tuning helped adapt it to the specific task. However, now we will use BERT only as a feature extractor. We will load a pretrained model, run the sentence through it, and use the pre-final layer to construct our features.

- In the previous chapter, we used TensorFlow for implementing BERT. Now, we will use PyTorch, a deep learning framework developed by researchers from Facebook. This provides a much more intuitive, straightforward, and understandable interface to design and run deep neural networks. It also has a `transformers` library, which provides easy implementations of all pretrained models.

First, we will initialize the BERT model and set it to evaluation mode. In the evaluation mode, there is no learning, just inferencing. Therefore, we need only the forward pass and no backpropagation:

```
import torch
from pytorch_transformers import BertTokenizer
from pytorch_transformers import BertModel

tokenizer = BertTokenizer.from_pretrained('bert-base-uncased')
model = BertModel.from_pretrained('bert-base uncased',
output_hidden_states=True)

model.eval()
```

We will now prepare our data in the format needed by BERT. This includes adding the two special tokens to indicate the start and separation. Then, we will run the model in inference mode to obtain the embeddings (hidden states). Recall that when we used Word2Vec embeddings, we averaged the embeddings for each word. In the case of BERT embeddings, we have multiple choices:

- Use just the last hidden state as the embedding:

```
X_train_BERT = []
for x in X_train:
  # Add CLS and SEP
  marked_text = "[CLS] " + x + " [SEP]"
```

```
# Split the sentence into tokens.
tokenized_text = tokenizer.tokenize(marked_text)
# Map the token strings to their vocabulary indices.
indexed_tokens = tokenizer.convert_tokens_to_ids(tokenized_
text)
tokens_tensor = torch.tensor([indexed_tokens])
with torch.no_grad():
  outputs = model(tokens_tensor)
  feature_vector = outputs[0]


X_train_BERT.append(feature_vector)
```

- Use the sum of all hidden states as the embedding:

```
X_train_BERT = []
for x in X_train:
  # Add CLS and SEP
  marked_text = "[CLS] " + x + " [SEP]"
  # Split the sentence into tokens.
  tokenized_text = tokenizer.tokenize(marked_text)
  # Map the token strings to their vocabulary indeces.
  indexed_tokens = tokenizer.convert_tokens_to_ids(tokenized_
text)
  tokens_tensor = torch.tensor([indexed_tokens])
  with torch.no_grad():
    outputs = model(tokens_tensor)
    hidden_states = outputs[2]
    feature_vector = torch.stack(hidden_states).sum(0)
  X_train_BERT.append(feature_vector)
```

- Use the sum of the last four layers as an embedding:

```
X_train_BERT = []
for x in X_train:
  # Add CLS and SEP
  marked_text = "[CLS] " + x + " [SEP]"
  # Split the sentence into tokens.
  tokenized_text = tokenizer.tokenize(marked_text)
  # Map the token strings to their vocabulary indeces.
  indexed_tokens = tokenizer.convert_tokens_to_ids(tokenized_
text)
  tokens_tensor = torch.tensor([indexed_tokens])
  with torch.no_grad():
    outputs = model(tokens_tensor)
    hidden_states = outputs[2]
```

```
        feature_vector = torch.stack(hidden_states[-4:]).sum(0)
    X_train_BERT.append(feature_vector)
```

- Concatenate the last four layers and use that as the embedding:

```
X_train_BERT = []
for x in X_train:
  # Add CLS and SEP
  marked_text = "[CLS] " + x + " [SEP]"
  # Split the sentence into tokens.
  tokenized_text = tokenizer.tokenize(marked_text)
  # Map the token strings to their vocabulary indeces.
  indexed_tokens = tokenizer.convert_tokens_to_ids(tokenized_
text)
  tokens_tensor = torch.tensor([indexed_tokens])
  with torch.no_grad():
    outputs = model(tokens_tensor)
    hidden_states = outputs[2]
    feature_vector = torch.cat([hidden_states[i] for i in [-1,-
2,-3,-4]], dim=-1)
  X_train_BERT.append(feature_vector)
```

Once we have the BERT features, we train and evaluate the model using our usual methodology. We will show an example of an MLP here, but the same process can be repeated for all the classifiers:

```
from sklearn.neural_network import MLPClassifier
model = MLPClassifier(hidden_layer_sizes = (1000, 700, 500, 200),max_
iter = 100,activation = 'relu',solver = 'adam',random_state = 123)
model.fit(X_train_BERT, Y_train)
Y_predicted = model.predict(X_test_BERT)
print(evaluate_model(Y_test, Y_pred))
```

This completes our analysis of how transformers can be used to detect machine-generated text.

Compare and contrast

By now, we have explored several techniques for detecting bot-generated news. Here's a list of all of them:

- Hand-crafted features such as function words, punctuation words, and automated readability index

- TF-IDF scores for words

- Word2Vec embeddings:

  - Averaged across the text for all words

  - Concatenated for each word across the text

- BERT embeddings:

  - Using only the last hidden state

  - Using the sum of all hidden states

  - Using the sum of the last four hidden states

  - Using the concatenation of the last four hidden states

We can see that we have eight feature sets at our disposal. Additionally, we experimented with four different models: random forests, logistic regression, SVM, and deep neural network (MLP). This means that we have a total of 32 configurations (feature set x model) that we can use for building a classifier to detect bot-generated fake news.

I leave it up to you to construct this 8x4 matrix and determine which is the best approach among all of them!

Summary

In this chapter, we described approaches and techniques for detecting bot-generated fake news. With the rising prowess of artificial intelligence and the widespread availability of language models, attackers are using automated text generation to run bots on social media. These sock-puppet accounts can generate real-looking responses, posts, and, as we saw, even news-style articles. Data scientists in the security space, particularly those working in the social media domain, will often be up against attackers who leverage AI to spew out text and carpet-bomb a platform.

This chapter aims to equip practitioners against such adversaries. We began by understanding how text generation exactly works and created our own dataset for machine learning experiments. We then used a variety of features (hand-crafted, TF-IDF, and word embeddings) to detect the bot-generated text. Finally, we used contextual embeddings to build improved mechanisms.

In the next chapter, we will study the problem of authorship attribution and obfuscation and the social and technical issues surrounding it.

7
Attributing Authorship and How to Evade It

The internet has provided the impetus to the fundamental right of freedom of expression by providing a public platform for individuals to voice their opinions, thoughts, findings, and concerns. Any person can express their views through an article, a blog post, or a video and post it online, free of charge in some cases (such as on Blogspot, Facebook, or YouTube). However, this has also led to malicious actors being able to generate misinformation, slander, libel, and abusive content freely. Authorship attribution is a task where we identify the author of a text based on the contents. Attributing authorship can help law enforcement authorities trace hate speech and threats to the perpetrator, or help social media companies detect coordinated attacks and Sybil accounts.

On the other hand, individuals may wish to remain anonymous as authors. They may want to protect their identity to avoid scrutiny or public interest. This is where authorship obfuscation comes into play. Authorship obfuscation is the task of modifying the text so that the author cannot be identified with attribution techniques.

In this chapter, we will cover the following main topics:

- Authorship attribution and obfuscation
- Techniques for authorship attribution
- Techniques for authorship obfuscation

By the end of this chapter, you will have an understanding of authorship attribution, the socio-technical aspects behind it, and methods to evade it.

Technical requirements

You can find the code files for this chapter on GitHub at https://github.com/PacktPublishing/10-Machine-Learning-Blueprints-You-Should-Know-for-Cybersecurity/tree/main/Chapter%207.

Authorship attribution and obfuscation

In this section, we will discuss exactly what authorship attribution is and the incentives for designing attribution systems. While there are some very good reasons for doing so, there are some nefarious ones as well; we will therefore also discuss the importance of obfuscation to protect against attacks by nefarious attackers.

What is authorship attribution?

Authorship attribution is the task of identifying the author of a given text. The fundamental idea behind attribution is that different authors have different styles of writing that will reflect in the vocabulary, grammar, structure, and overall organization of the text. Attribution can be based on heuristic methods (such as similarity, common word analysis, or manual expert analysis). Recent advances in **machine learning** (**ML**) have also made it possible to build classifiers that can learn to detect the author of a given text.

Authorship attribution is not a new problem—the study of this field goes back to 1964. A series of papers known as *The Federalist Papers* had been published, which contained over 140 political essays. While the work was jointly authored by 3 people, 12 of those essays were claimed by 2 authors. The study by Mosteller and Wallace involving Bayesian modeling and statistical analysis using n-grams, which produced statistically significant differences between the authors, is known to be the first actual work in authorship attribution.

Authorship attribution is important for several reasons, as outlined here:

- **Historical significance**: Scientists and researchers rely on historical documents and texts for evidence of certain events. At times, these may have immense political and cultural significance, and knowing the author would help place them in the proper context and determine their credibility. For example, if an account describing certain historical periods and projecting dictators or known malicious actors in a positive light were to be found, it would be important to ascertain who the author is, as that could change the credibility of the text. Authorship attribution would help in determining whether the text could be accepted as an authoritative source or not.

- **Intellectual property**: As with *The Federalist Papers*, there is often contention on who the owner of certain creative or academic works is. This happens when multiple people claim ownership over the same book, article, or research paper. At other times, one individual may be accused of plagiarizing the work of another. In such cases, it is extremely important to trace who the author of a particular text is. Authorship attribution can help identify the author, match similarity in style and tone, and resolve issues of contended intellectual property.

- **Criminal investigation**: Criminals often use text as a means of communicating with victims and law enforcement. This can be in the form of a ransom note or threats. If there is a significant amount of text, it may be possible that it reflects some of the stylistic habits of the author. Law enforcement officers use authorship attribution methods to determine whether the messages received fit the style of any known criminal.

- **Abuse detection**: Sybil accounts are a growing challenge on the internet and social media. These are a group of accounts controlled by the same entity but masquerading as different people. Sybil accounts have nefarious purposes such as multiple Facebook accounts generating fake engagement, or multiple Amazon accounts to write fake product reviews. As they are controlled by the same entity, the content produced (posts, tweets, reviews) is generally similar. Authorship attribution can be used to identify groups of accounts that post content written by the same author.

With the prevalence of the internet and social media platforms, cybercrime has been on the rise, and malicious actors are preying on unknowing victims. Authorship attribution, therefore, is also a cybersecurity problem. The next section will describe authorship obfuscation, a task that counters authorship attribution.

What is authorship obfuscation?

In the previous section, we discussed authorship attribution, which is the task of identifying the author of a given text. Authorship obfuscation is a task that works exactly toward the opposite.

Given a text, authorship obfuscation aims to manipulate and modify the text in such a way that the end result is this:

- The meaning and key points in the text are left intact
- The style, structure, and vocabulary are suitably modified so that the text cannot be attributed to the original author (that is, authorship attribution techniques will be evaded)

Individuals may use obfuscation techniques to hide their identity. Consider the sentence *"We have observed great corruption at the highest levels of government in this country."* If this is re-written as *"Analysis has shown tremendous corrupt happenings in the uppermost echelons of this nation's administration,"* the meaning is left intact. However, the style is clearly different and does not bear much resemblance to the original author. This is effective obfuscation. An analyst examining the text will not be easily able to map it to the same author.

Note that both of the objectives in obfuscation (that is, retaining the original meaning and stripping off the style markers) are equally important and there is a trade-off between them. We can obtain high obfuscation by making extreme changes to the text, but at that point, the text may have lost its original meaning and intent. On the other hand, we can retain the meaning with extremely minor tweaks—but this may not lead to effective obfuscation.

Authorship obfuscation has both positive and negative use cases. Malicious actors can use obfuscation techniques in order to counter the attribution purposes discussed previously and avoid detection. For example, a criminal who wants to stay undetected and yet send ransom notes and emails may obfuscate their text by choosing a different vocabulary, grammatical structure, and organization. However, obfuscation has several important use cases in civil and human rights, as detailed here:

- **Oppressive governments**: As discussed before, the internet has greatly facilitated the human right to freely express oneself. However, some governments may try to curtail these rights by targeting individuals who speak up against them. For example, an autocratic government may want to prohibit reporting on content that speaks against its agenda or expose corruption and malicious schemes. At such times, journalists and individuals may want to remain anonymous—their identity being detected could lead to them being captured. Obfuscation techniques will alter the text they write so that the matter they want to convey will be retained, but the writing style will be significantly different than their usual one.

- **Sensitive issues**: Even if the government is not oppressive by nature, certain issues may be sensitive to discuss and controversial. Examples of such issues include religion, racial discrimination, reports of sexual violence, homosexuality, and reproductive healthcare. Individuals who write about such issues may offend the public or certain other groups or sects. Authorship obfuscation allows such individuals to publish such content and yet remain anonymous (or, at least, makes it harder to discern the author of the text).

- **Privacy and anonymity**: Many believe that privacy is a fundamental human right. Therefore, even if an issue is not sensitive or the government is not corrupt, users have the right to protect their identity if they want to. Every individual should be free to post what they want and hide their identity. Authorship obfuscation allows users to maintain their privacy while expressing themselves.

Now that you have a good understanding of authorship attribution and obfuscation and why it is actually needed, let us go into implementing it with Python.

Techniques for authorship attribution

The previous section described the importance of authorship attribution and obfuscation. This section will focus on the attribution aspect—how we can design and build models to pinpoint the author of a given text.

Dataset

There has been prior research in the field of authorship attribution and obfuscation. The standard dataset for benchmarking on this task is the *Brennan-Greenstadt Corpus*. This dataset was collected through a survey at a university in the United States. 12 authors were recruited, and each author was required to submit a pre-written text that comprised at least 5,000 words.

A modified and improved version of this data—called the *Extended Brennan-Greenstadt Corpus*—was released later by the same authors. To generate this dataset, the authors conducted a large-scale survey by recruiting participants from Amazon **Mechanical Turk (MTurk)**. MTurk is a platform that allows researchers and scientists to conduct human-subjects research. Users sign up for MTurk and fill out detailed questionnaires, which makes it easier for researchers to survey the segment or demographic (by gender, age, nationality) they want. Participants get paid for every **human interaction task (HIT)** they complete.

To create the extended corpus, MTurk was used so that the submissions would be diverse and varied and not limited to university students. Each piece of writing was scientific or scholarly (such as an essay, a research paper, or an opinion paper). The submission only contained text and no other information (such as references, citations, URLs, images, footnotes, endnotes, and section breaks). Quotations were to be kept to a minimum as most of the text was supposed to be author generated. Each sample had at least 500 words.

Both the *Brennan-Greenstadt Corpus* and the *Extended Brennan-Greenstadt Corpus* are available online to the public for free. For simplicity, we will run our experiments with the *Brennan-Greenstadt Corpus* (which contains writing samples from university students). However, readers are encouraged to reproduce the results on the extended corpus, and tune models as required. The process and code would remain the same—you would have to just change the underlying dataset.

For convenience, we have provided the dataset we're using (`https://github.com/PacktPublishing/10-Machine-Learning-Blueprints-You-Should-Know-for-Cybersecurity/blob/main/Chapter%207/Chapter_7.ipynb`). The dataset consists of a root folder that has one subfolder for every author. Each subfolder contains writing samples for the author. You will need to unzip the data and place it into the folder you want (and change `data_root_dir` in the following code accordingly).

Recall that for our experiments, we need to read the dataset such that the input (features) is in an array and the labels are in a separate array. The following code snippet parses the folder structure and produces data in this format:

```
def read_dataset(num_authors = 99):
  X = []
  y = []

  data_root_dir = "../data/corpora/amt/"
  authors_to_ignore = []
  authorCount = 0

  for author_name in os.listdir(data_root_dir):
      # Check if the maximum number of authors has been parsed
      if authorCount > self.numAuthors:
        break
```

```
        if author_name not in authors_to_ignore:
            label = author_name
            documents_path = data_root_dir + author_name + "/"
            authorCount += 1

            for doc in os.listdir(documents_path):
                if validate_file(doc):
                    text = open(docPath + doc, errors = "ignore").read()
                    X.append(text)
                    y.append(label)

    return X, y
```

The dataset also contains some housekeeping files as well as some files that indicate the training, test, and validation data. We need a function to filter out these so that this information is not read in the data. Here's what we'll use:

```
def validate_file(file_name):
    filterWords = ["imitation", "demographics", "obfuscation",
"verification"]
    for fw in filterWords:
        if fw in file_name:
            return False
    return True
```

Our dataset has been read, and we can now extract features from it. For authorship attribution, most features are stylometric and hand-crafted. In the next section, we will explore some features that have shown success in prior work.

Feature extraction

We will now implement a series of functions, each of which extracts a particular feature from our data. Each function will take in the input text as a parameter, process it, and return the feature as output.

We begin as usual by importing the required libraries:

```
import os
import nltk
import re
import spacy
from sortedcontainers import SortedDict
from keras.preprocessing import text
import numpy as np
```

As a first feature, we will use the number of characters in the input:

```
def CountChars(input):
    num_chars = len(input)
    return num_chars
```

Next, we will design a feature that measures the average word length (number of characters per word). For this, we first split the text into an array of words and clean it up by removing any special characters such as braces, symbols, and punctuation. Then, we calculate the number of characters and the number of words separately. Their ratio is our desired feature:

```
def averageCharacterPerWord(input):
    text_array = text.text_to_word_sequence(input,
                                            filters=' !#$%&()*+,-
./:;<=>?@[\\]^_{|}~\t\n"',
                                            lower=False, split=" ")
    num_words = len(text_array)

    text_without_spaces = input.replace(" ", "")
    num_chars = len(text_without_spaces)

    avgCharPerWord = 1.0 * num_chars / num_words
    return avgCharPerWord
```

Now, we calculate the frequency of alphabets. We will first create a 26-element array where each element counts the number of times that alphabet appears in the text. The first element corresponds to A, the next to B, and so on. Note that as we are counting alphabets, we need to convert the text to lowercase. However, if this were our feature, it would depend heavily on the length of the text. Therefore, we normalize this by the total number of characters. Each element of the array, therefore, depicts the percentage of that particular alphabet in the text:

```
def frequencyOfLetters(input):
    input = input.lower()  # because its case sensitive
    input = input.lower().replace(" ", "")
    num_chars = len(input)

    characters = "abcdefghijklmnopqrstuvwxyz".split()
    frequencies = []

    for each_char in characters:
        char_count = input.count(each_char)
        if char_count < 0:
            frequencies.append(0)
        else:
```

```
        frequencies.append(char_count/num_chars)

    return frequencies
```

Next, we will calculate the frequency of common bigrams. Prior research in linguistics and phonetics has indicated which bigrams are common in English writing. We will first compile a list of such bigrams. Then, we will parse through the list and calculate the frequency of each bigram and compute a vector. Finally, we normalize this vector, and the result represents our feature:

```
def CommonLetterBigramFrequency(input):

    common_bigrams = ['th','he','in','er','an','re','nd',
                      'at','on','nt','ha','es','st','en',
                      'ed','to','it','ou','ea','hi','is',
                      'or','ti','as','te','et','ng','of',
                      'al','de','se','le','sa','si','ar',
                      've','ra','ld','ur']
    bigramCounter = []

    input = input.lower().replace(" ", "")

    for bigram in common_bigrams:
      bigram_count = input.count(bigram)
      if bigram_count == -1:
        bigramCounter.append(0)
      else:
        bigramCounter.append(bigram_count)

    total_bigram_count = np.sum(bigramCounter)
    bigramCounterNormalized = []
    for bigram_count in bigramCounter:
      bigramCounterNormalized.append(bigram_count / total_bigram_
count)

    return bigramCounterNormalized
```

Just as with the common bigrams, we also compute the frequency of common trigrams (sequences of three alphabets). The final feature represents a normalized vector, similar to what we had for bigrams:

```
def CommonLetterTrigramFrequency(input):

    common_trigrams = ["the", "and", "ing", "her", "hat",
                       "his", "tha", "ere", "for", "ent",
                       "ion", "ter", "was", "you", "ith",
```

```
                    "ver", "all", "wit", "thi", "tio"]
    trigramCounter = []

    input = input.lower().replace(" ", "")

    for trigram in common_trigrams:
      trigram_count = input.count(trigram)
      if trigram_count == -1:
        trigramCounter.append(0)
      else:
        trigramCounter.append(trigram_count)

    total_trigram_count = np.sum(trigramCounter)
    trigramCounterNormalized = []
    for trigram_count in trigramCounter:
      trigramCounterNormalized.append(trigram_count / total_trigram_
  count)

    return trigramCounterNormalized
```

The next feature is the percentage of characters that are digits. First, we calculate the total number of characters in the text. Then, we parse through the text character by character and check whether each character is numeric. We count all such occurrences and divide them by the total number we computed earlier—this gives us our feature:

```
def digitsPercentage(input):

    num_chars = len(input)
    num_digits = 0

    for each_char in input:
      if each_char.isnumeric():
        num_digits = num_digits + 1

    digit_percent = num_digits / num_chars
    return digit_percent
```

Similarly, the next feature is the percentage of characters that are alphabets. We will first need to convert the text to lowercase. Just as with the previous feature, we parse character by character, now checking whether each character we encounter is in the range [a-z]:

```
def charactersPercentage(input):

    input = input.lower().replace(" ", "")
```

```
characters = "abcdefghijklmnopqrstuvwxyz"

total_chars = len(input)
char_count = 0

for each_char in input:
  if each_char in characters:
    char_count = char_count + 1

char_percent = char_count / total_chars
return char_percent
```

Previously, we calculated the frequency of alphabets. On similar lines, we calculate the frequency of each digit from 0 to 9 and normalize it. The normalized vector is used as our feature:

```
def frequencyOfDigits(input):

    input = input.lower().replace(" ", "")
    num_chars = len(input)

    digits = "0123456789".split()
    frequencies = []

    for each_digit in digits:
      digit_count = input.count(each_digit)
      if digit_count < 0:
        frequencies.append(0)
      else:
        frequencies.append(digit_count/num_chars)

    return frequencies
```

We will now calculate the percentage of characters that are uppercase. We follow a similar procedure as we did for counting the characters, but now we count for capital letters instead. The result is normalized, and the normalized value forms our feature:

```
def upperCaseCharactersPercentage(input):

    input = input.replace(" ", "")
    upper_characters = "ABCDEFGHIJKLMNOPQRSTUVWXYZ"

    num_chars = len(input)
    upper_count = 0
```

```
for each_char in upper_characters:
  char_count = input.count(each_char)
  if char_count > 0:
    upper_count = upper_count + char_count

upper_percent = upper_count / num_chars
return upper_percent
```

Now, we will calculate the frequency of special characters in our text. We first compile a list of special characters of interest in a file. We parse the file and count the frequency of each character and form a vector. Finally, we normalize this vector by the total number of characters. Note that the following function uses a static file where the list of characters is stored—you will need to change this line of code to reflect the path where the file is stored on your system:

```
def frequencyOfSpecialCharacters(input):

    SPECIAL_CHARS_FILE = "static_files/writeprints_special_chars.txt"
    num_chars = len(input)
    special_counts = []

    special_characters = open(SPECIAL_CHARS_FILE , "r").readlines()
    for each_char in special_characters:
      special = each_char.strip().rstrip()
      special_count = input.count(special)
      if special_count < 0:
        special_counts.append(0)
      else:
        special_counts.append(special_count / num_chars)

    return special_counts
```

Next, we will count the number of short words in the text. We define a short word as one with fewer than or at most three characters. This is a rather heuristic definition; there is no globally accepted standard for a word being short. You can play around with different values here and see whether it affects the results:

```
def CountShortWords(input):
    words = text.text_to_word_sequence(input,
filters=",.?!\"'`;:-()&$", lower=True, split=" ")
    short_word_count = 0

    for word in words:
        if len(word) <= 3:
            short_word_count = short_word_count + 1

    return short_word_count
```

As a very simple feature, we compute the total number of words in the input. This involves splitting the text into an array of words (cleaning up special characters) and counting the length of the array:

```
def CountWords(input):
    words = text.text_to_word_sequence(input,
filters=",.?!\"'`;:-()&$", lower=True, split=" ")
    return len(words)
```

Now, we calculate the average word length. We simply calculate the length of each word in the text and use the mean of all such length values as the feature:

```
def averageWordLength(input):
    words = text.text_to_word_sequence(inputText,
filters=",.?!\"'`;:-()&$", lower=True, split=" ")
    lengths = []
    for word in words:
        lengths.append(len(word))
    return np.mean(lengths)
```

We now have all of the functions to compute features in place. Each function will take in the text as a parameter and process it to produce the feature we designed. Now, we will write a wrapper function to put it all together. This function, on being passed the text, will run it through all of our feature extraction functions and compute each feature. Each feature will be appended to a vector. This forms our final feature vector:

```
def calculate_features(input):

    features = []

    features.extend([CountWords(input)])
    features.extend([averageWordLength(input)])
    features.extend([CountShortWords(input)])
    features.extend([CountChars(input)])
    features.extend([averageCharacterPerWord(input)])
    features.extend([frequencyOfLetters(input)])
    features.extend([CommonLetterBigramFrequency(input)])
    features.extend([CommonLetterTrigramFrequency(input)])
    features.extend([digitsPercentage(input)])
    features.extend([charactersPercentage(input)])
    features.extend([frequencyOfDigits(input)])
    features.extend([upperCaseCharactersPercentage(input)])
    features.extend([frequencyOfSpecialCharacters(input)])
    features.extend([frequencyOfPunctuationCharacters(input)])
    features.extend([posTagFrequency(input)])
```

Now, all that is to be done is to apply this function to our dataset:

```
X_original, Y = read_dataset(num_authors = 6)
X_Features = []
for x in X_original:
  x_features = calculate_features(x)
  X.append(x_features)
```

After this is executed, X will be an array containing features that we designed, and Y will contain the corresponding labels. The hard part is done! Next, we will turn to the modeling phase.

Training the attributor

In the previous section, we processed our dataset, hand-crafted several features, and now have one feature vector per text and the ground-truth label corresponding to it. At this point, this is essentially a **supervised learning (SL)** problem; we have the features and labels and want to learn the association between them. We will approach this as we did with all other supervised problems we have seen so far.

To recap, here are the steps we'll take:

1. Split the data into training and testing sets.
2. Train a supervised classifier on the training set.
3. Evaluate the performance of the trained model on the testing set.

First, we split the data as follows. Note that we have a mix of authors, therefore we have multiple labels. We must ensure that the distribution of labels in the training and test sets is roughly similar; otherwise, our model will be biased toward specific authors. If a particular author does not appear in the training set, the model will not be able to detect them at all.

Then, we train our classification model (logistic regression, decision tree, random forest, **deep neural network (DNN)**) on the training set. We use this model to make predictions for the data in the test set and compare the predictions with the ground truth. As this procedure has been covered in preceding chapters, we will not go into detailed explanations here.

A sample code snippet that performs the previous steps with a random forest is shown next. Readers should repeat it with other models as well:

```
# Import Packages
from sklearn.model_selection import train_test_split
from sklearn.ensemble import RandomForestClassifier
from sklearn.metrics import confusion_matrix
from matplotlib import pyplot as plt
import seaborn as sns
```

```
# Training and Test Datasets
X_train, X_test, Y_train, Y_test = train_test_split(X_Features, Y)

# Train the model
model = RandomForestClassifier(n_estimators = 100)
model.fit(X_train, Y_train)

# Plot the confusion matrix
Y_predicted = model.predict(X_test)
confusion = confusion_matrix(Y_test, Y_predicted)
plt.figure(figsize = (10,8))
sns.heatmap(confusion, annot = True,
            fmt = 'd', cmap="YlGnBu")
```

As you run this, you will notice that the confusion matrix now looks different. Whereas previously we had a 2x2 matrix, now we get a 6x6 matrix. This is because our dataset now contains six different labels (one for every author). Therefore, for every data point with a given class, there are six possible classes to be predicted.

Calculating accuracy is still the same; we need to find the fraction of examples that were predicted correctly. Here is a function that does this:

```
def calculate_accuracy(actual, predicted):
  total_examples = len(actual)
  correct_examples = 0

  for idx in range(total_examples):
    if actual[i] == predicted[i]:
      correct_examples = correct_examples + 1

  accuracy = correct_examples / total_examples
  return accuracy
```

In multi-class problems, the definitions of precision and recall are no longer as simple as computing false positives and negatives. Rather, these metrics are calculated per class. For example, if there are six labels (1-6), then for class 2, we say the following:

- True positives are those where the actual and predicted classes are both 2

- False positives are the ones where the predicted class is 2, but the actual class is anything other than 2

- True negatives are those where both the actual and predicted classes are anything other than 2

- False negatives are those where the predicted class is anything other than 2, but the actual class is 2

Using these definitions and the usual expressions for calculating metrics, we can calculate per-class metrics. The per-class precision and recall may be averaged to compute the overall precision, recall, and F1 scores.

Fortunately, we do not need to manually implement this per-class metric calculation. `scikit-learn` has an inbuilt classification report that will compute and produce these metrics for you. This can be used as follows:

```
from sklearn.metrics import classification_report
classification_report(Y_test, Y_predicted)
```

This completes our implementation and analysis of authorship attribution. Next, we will suggest some experiments that readers can pursue to explore the topic more.

Improving authorship attribution

We have presented vanilla models and techniques for authorship attribution. However, there is a large scope for improvement here. As data scientists, we must be willing to explore new ideas and techniques and continuously improve our models. Here are a few suggestions that readers should try out to see whether they can obtain a better performance.

Additional features

We have used the feature set that is known as the Writeprints set of features. This has shown success in prior research. However, this is not an exhaustive list of features. Readers can explore more hand-crafted and automatic features to evaluate whether performance is improved. Examples of some features are set out here:

- Text sentiment
- Text polarity
- Number and fraction of function words
- **Term Frequency – Inverse Document Frequency (TF-IDF)** features
- Word embeddings derived from Word2vec
- Contextual word embeddings derived from **Bidirectional Encoder Representations from Transformers (BERT)**

Data configurations

The experiment we ran was on a subset of six authors from the dataset. In the real world, the problem is much more open-ended and there may be several more authors. It is worth exploring how the model performance varies as the number of authors changes. In particular, readers should explore the following:

- What are the performance measures if we choose only 3 authors? What about if we choose 12?

- How does the performance change if we model this problem as binary classification? Instead of predicting the author, we predict whether a particular text was written by a particular author or not. This would involve training a separate classifier per author. Does this show better predictive power and practical application than the multi-class approach?

Model improvements

For brevity and to avoid repetition, we showed only the example of a random forest. However, readers should experiment with more models, including but not limited to the following:

- **Support vector machines (SVMs)**
- Naïve-Bayes classifier
- Logistic regression
- Decision tree
- DNN

The **neural network (NN)** algorithms will be particularly useful as the number of features increases. When the embeddings and TF-IDF scores are added, the features will not be easily interpretable anymore—NNs excel in such situations where they can discover high-dimensional features.

This completes our discussion of authorship attribution. In the next section, we will discuss a problem that is the opposite of the attribution task.

Techniques for authorship obfuscation

So far, we have seen how authorship can be attributed to the writer and how to build models to detect the author. In this section, we will turn to the authorship obfuscation problem. Authorship obfuscation, as discussed in the initial section of this chapter, is the art of purposefully manipulating the text to strip it of any stylistic features that might give away the author.

The code is inspired by an implementation that is freely available online (`https://github.com/asad1996172/Obfuscation-Systems`) with a few minor tweaks.

First, we will import the required libraries. The most important library here is the **Natural Language Toolkit (NLTK)** library (`https://www.nltk.org/`) developed by Stanford. This library contains standard off-the-shelf implementations for several **natural language processing (NLP)** tasks such as tokenization, **part-of-speech (POS)** tagging, **named entity recognition (NER)**, and so on. It has a powerful set of functionalities that greatly simplify feature extraction in text data. You are encouraged to explore the library in detail. The **word-sense disambiguation (WSD)** implementation (`https://github.com/asad1996172/Obfuscation-Systems/blob/master/Document%20Simplification%20PAN17/WSD_with_UKB.py`) can be found online and should be downloaded locally.

The code to import the libraries is shown here:

```
import nltk
import re
import random
import pickle
from nltk.wsd import lesk
from nltk.corpus import wordnet as wn
import WSD_with_UKB as wsd
from nltk.tokenize import sent_tokenize
from nltk.tokenize import RegexpTokenizer
```

First, we will implement a function for the expansion and contraction replacement. We begin by reading the extraction-contraction list from the `pickle` file (you will have to change the path to it accordingly). The result is a dictionary where the keys are contractions and values associated are corresponding expansions. We parse through the sentence and count the expansions and contractions occurring. If there are mostly contractions, we replace them with expansions, and if there are mostly expansions, we replace them with contractions. If both are the same, we do nothing at all:

```
def contraction_replacement(sentence):

    # Read Contractions
    CONTRACTION_FILE = 'contraction_extraction.pickle'
    with open(CONTRACTION_FILE, 'rb') as contraction_file:
        contractions = pickle.load(contraction_file)

    # Calculate contraction counts
    all_contractions = contractions.keys()
    contractions_count = 0
    for contraction in all_contractions:
        if contraction.lower() in sentence.lower():
            contractions_count += 1

    # Calculate expansion counts
    all_expansions = contractions.values()
    expansions_count = 0
    for expansion in all_expansions:
        if expansion.lower() in sentence.lower():
            expansions_count += 1

    if contractions_count > expansions_count:
        # There are more contractions than expansions
        # So we should replace all contractions with their expansions
```

```
        temp_contractions = dict((k.lower(), v) for k, v in
contractions.items())
        for contraction in all_contractions:
            if contraction.lower() in sentence.lower():
                case_insensitive = re.compile(re.escape(contraction.
lower()), re.IGNORECASE)
                sentence = case_insensitive.sub(temp_
contractions[contraction.lower()], sentence)
        contractions_applied = True

    elif expansions_count > contractions_count:
        # There are more expansions than contractions
        # So we should replace expansions by contractions
        inv_map = {v: k for k, v in contractions.items()}
        temp_contractions = dict((k.lower(), v) for k, v in inv_map.
items())
        for expansion in all_expansions:
            if expansion.lower() in sentence.lower():
                case_insensitive = re.compile(re.escape(expansion.
lower()), re.IGNORECASE)
                sentence = case_insensitive.sub(temp_
contractions[expansion.lower()], sentence)
        contractions_applied = True

    else:
        # Both expansions and contractions are equal
        # So do nothing
        contractions_applied = False

    return sentence, contractions_applied
```

Next, we will remove any parentheses occurring in the text. This means that we have to search for characters associated with brackets— (,) , [,] , {, } —and remove them from the text:

```
def remove_parenthesis(sentence):
    parantheses = ['(', ')', '{', '}', '[', ']']
    for paranthesis in parantheses:
        sentence = sentence.replace(paranthesis, "")
    return sentence
```

We will implement a function to purge discourse markers from the text. We will first read a list of discourse markers (you will need to change the filename and path, depending on how you have saved it locally). We then iterate through the list and remove each item from the text, if found:

```
def remove_discourse_markers(sentence):

    # Read Discourse Markers
```

```
        DISCOURSE_FILE = 'discourse_markers.pkl'
        with open(DISCOURSE_FILE , 'rb') as discourse_file:
            discourse_markers = pickle.load(discourse_file)

        sent_tokens = sentence.lower().split()
        for marker in discourse_markers:
            if marker.lower() in sent_tokens:
                case_insensitive = re.compile(re.escape(marker.lower()),
re.IGNORECASE)
                sentence = case_insensitive.sub('', sentence)

        return sentence
```

Next, we will implement a function to remove appositions from the text. We will use **regular expression (regex)** matching for this:

```
def remove_appositions(sentence):
    sentence = re.sub(r" ?\,[(^)]+\,", "", sentence)
    return sentence
```

We will now implement a function to change expressions of possession. We will first use regex matching to find expressions of the form "X of Y." We will then replace this with "Y's X." For example, "book of Jacob" will become "Jacob's book." Note that we are not making this replacement deterministically. We will randomly choose whether to replace or not (biased with the probability of replacement being 2/3):

```
def apply_possessive_transformation(text):
    if re.match(r"(\w+) of (\w+)", text):
        rnd = random.choice([False, True, False])
        if rnd:
            return re.sub(r"(\w+) of (\w+)" , r"\2's \1", text)
    return text
```

Next, we will apply equation transformation where we will replace mathematical expressions with their textual representations. We will define a dictionary where common symbols and their text representations are defined (such as "+" translating to "plus" and "*" translating to "multiplied by"). Then, we will find occurrences of each symbol in the text and make the necessary replacements:

```
def apply_equation_transformation(text):
    words = RegexpTokenizer(r'\w+').tokenize(text)
    symbol_to_text =   {
                '+': ' plus ',
                '-': ' minus ',
                '*': ' multiplied by ',
                '/': ' divided by ',
```

```
                   '=': ' equals ',
                   '>': ' greater than ',
                   '<': ' less than ',
                   '<=': ' less than or equal to ',
                   '>=': ' greater than or equal to ',
            }
    for n,w in enumerate(words):
        for symbol in symbol_to_text:
            if symbol in w:
                words[n] = words[n].replace(symbol, symbol_to_
text[sym])

    sentence = ''
    for word in words:
      sentence = sentence + word + " "

    return sentence
```

The next step is synonym replacement. However, as a helper function for synonym replacement, we need a function for *untokenization*. This is the exact opposite of tokenization and can be done with the following code:

```
def untokenize(words):
    text = ' '.join(words)
    step1 = text.replace("`` ", '"').replace(" ''", '"').replace('. .
. ', '...')
    step2 = step1.replace(" ( ", " (").replace(" ) ", ") ")
    step3 = re.sub(r' ([.,:;?!%]+)([ \'"`])', r"\1\2", step2)
    step4 = re.sub(r' ([.,:;?!%]+)$', r"\1", step3)
    step5 = step4.replace(" '", "'").replace(" n't", "n't").replace(
        "can not", "cannot")
    step6 = step5.replace(" ` ", " '")
    return step6.strip()
```

Now, we will implement the actual synonym substitution:

```
def synonym_substitution(sentence, all_words):
    new_tokens = []
    output = wsd.process_text(sentence)

    for token, synset in output:
        if synset != None:
            try:
                # Get the synset name
                synset = synset.split('-')
```

```
            offset = int(synset[0])
            pos = synset[1]
            synset_name = wn.synset_from_pos_and_offset(pos,
offset)


            # List of Synonyms
            synonyms = synset_name.lemma_names()


            for synonym in synonyms:
                if synonym.lower() not in all_words:
                    token = synonym
                    break

        except Exception as e:
            # Some error in the synset naming....
            continue


    new_tokens.append(token)

    final = untokenize(new_tokens)
    final = final.capitalize()
    return final
```

Finally, we will put all of this together in a wrapper function. This function will take in the text and apply all of our transformations (contraction-expansion replacement, parenthesis removal, discourse and apposition removal, synonym replacement, equation transformation, and possessive transformation) to each sentence of the text, and then join the sentences back to form the obfuscated text:

```
def obfuscate_text(input_text):
    obfuscated_text = []
    sentences = sent_tokenize(input_text)
    tokens = set(nltk.word_tokenize(input_text.lower()))


    for sentence in sentences:
        # 1. Apply Contractions
        sentence, contractions_applied = contraction_
replacement(sentence, contractions)


        # 2. Remove Parantheses
        sentence = remove_parenthesis(sentence)


        # 3. Remove Discourse Markers
```

```
            sentence = remove_discourse_markers(sentence, discourse_
markers)

        # 4. Remove Appositions
        sentence = remove_appositions(sentence)

        # 5. Synonym Substitution
        sentence = synonym_substitution(sentence, tokens)

        # 6. Apply possessive transformation
        sentence = apply_possessive_transformation(sentence)

        # 7. Apply equation transformation
        sentence = apply_equation_transformation(sentence)

        obfuscated_text.append(sentence)

    obfuscated_text = " ".join(obfuscated_text)
    return obfuscated_text
```

We now will test how effective this obfuscation is. We will train a vanilla model and then test it on the obfuscated data. This mirrors exactly the threat model that would occur in the real world; at the time of training, we would not have access to the obfuscated data. Here is the process we will follow in order to evaluate the model:

1. Split the data into training and test sets.
2. Extract features from the training data.
3. Train an authorship attribution ML model based on these features.
4. Apply the obfuscator on the test data to transform the raw text into obfuscated text.
5. Extract features from the obfuscated text and use them to run inference on the previously trained model.

We load the data and split it as before:

```
from sklearn.model_selection import train_test_split

# Read Data
X, Y = read_dataset(num_authors = 6)

# Split it into train and test
X_train, X_test, Y_train, Y_test = train_test_split(X, Y)
```

Then, we extract features and train a model. Note that we extract features only from the training data, not the test data (which we need to obfuscate):

```
# Extract features from training data
X_train_features = []
for x in X_train:
    x_features = calculate_features(x)
    X_train_features.append(x_features)

# Train the model
model = RandomForestClassifier(n_estimators = 100)
model.fit(X_train_features, Y_train)
```

Now, we will obfuscate the test data using the functions we defined earlier, and then extract features from the obfuscated version of the data:

```
X_test_obfuscated = []
for x in X_test:
    # Obfuscate
    x_obfuscated = obfuscate_text(x)
    # Extract features
    x_obfuscated_features = calculate_features(x_obfuscated)

    X_test_obfuscated.append(x_obfuscated_features)
```

Finally, we can run inference on the trained model using the newly generated (obfuscated) data:

```
# Calculate accuracy on original
Y_pred_original = model.predict(X_test)
accuracy_orig = calculate_accuracy(Y_test, Y_pred_original)

# Calculate accuracy on obfuscated
Y_pred_obfuscated = model.predict(X_test_obfuscated)
accuracy_obf = calculate_accuracy(Y_test, Y_pred_obfuscated)
```

Comparing the two values of accuracy should give you the performance degradation caused due to the obfuscation. The first calculated value represents the accuracy of the original data, and the second one represents the accuracy of the model when our obfuscation tactics are applied. When the second value is lower than the first, our obfuscation has been successful.

Next, we will provide an overview of some strategies to improve the performance of our obfuscators.

Improving obfuscation techniques

Here, we describe potential changes and improvements that can help us achieve a better performance of our obfuscator. Readers are highly encouraged to experiment with these to examine which ones show the best performance.

Advanced manipulations

In our example obfuscator, we implemented basic obfuscation tactics such as the replacement of synonyms, changing of contractions, removing parentheses, and so on. There is a vast arena of features that can be manipulated here. A few possibilities are given next:

- **Antonym replacement**: Replacing words with the negation of their antonyms. For example, *good* is replaced by *not bad*.

- **Function word manipulation**: Adding extra helper words at the beginning of sentences, or removing existing words that add no value. For example, *"Thus, we have shown that the plan works"* becomes *"We have shown that the plan works."*

- **Punctuation manipulation**: Adding punctuation symbols (two question marks, two exclamation marks, trailing periods) or removing existing ones. This may affect the grammar and structure of the sentence, which may or may not be acceptable depending on your use case.

Language models

Recent advances such as transformers and the attention mechanism have led to the development of several improved language models, which have excellent text-generation capabilities. Such models can be used to generate obfuscated text. An example is using a transformer-based document summarizer as an obfuscator. The summarizer aims to reproduce the text in the original document in a short and concise manner. The hope is that in doing so, it will strip off the stylistic features from the text. Readers are encouraged to experiment with various summarization models and compare the accuracy before and after obfuscation. Note that it is also important to check the similarity of the text against the original in terms of meaning.

This completes our discussion of authorship obfuscation models!

Summary

This chapter focused on two important problems in security and privacy. We began by discussing authorship attribution, a task of identifying who wrote a particular piece of text. We designed a series of linguistic and text-based features and trained ML models for authorship attribution. Then, we turned to authorship obfuscation, a task that aims to evade the attribution models by making changes to the text such that author-identifying characteristics and style markers are removed. We looked at a series of obfuscation methods for this. For both tasks, we looked at the improvements that could be made to the performance.

Both authorship attribution and obfuscation have important applications in cybersecurity. Attribution can be used to detect Sybil accounts, trace cybercriminals, and protect intellectual property rights. Similarly, obfuscation can help preserve the anonymity of individuals and provide privacy guarantees. This chapter enables ML practitioners in cybersecurity and privacy to effectively tackle these two tasks.

In the next one, we will change tracks slightly and look at how fake news can be detected using graph ML.

8

Detecting Fake News with Graph Neural Networks

In the previous chapters, we looked at tabular data, which was comprised of individual data points with their own features. While modeling and running our experiments, we did not consider any features of the relationship among the data points. Much real-world data, particularly that in the domain of cybersecurity, can naturally occur as graphs and be represented as a set of nodes, some of which are connected using edges. Examples include social networks, where users, photos, and posts can be connected using edges. Another example is the internet, which is a large graph of computers connected to each other.

Traditional machine learning algorithms cannot directly learn from graphs. Algorithms such as regression, neural networks, and trees, and optimization techniques such as gradient descent are designed to operate on Euclidean (flat) data structures. This has led to the development of **Graph Neural Networks** (**GNNs**), an upcoming area of research in the field of machine learning. This has found tremendous applications in cybersecurity, particularly in areas such as botnets, fake news detection, and fraud analytics. This chapter will focus on detecting fake news using GNNs. We will first cover the basics of graph theory, followed by how graph machine learning can be used as a tool to frame a security problem. Although we will learn how to use GNN models to detect fake news, the techniques we will introduce are generic and can be applied to multiple problems.

In this chapter, we will cover the following main topics:

- An introduction to graphs
- Machine learning on graphs
- Fake news detection with GNNs

By the end of this chapter, you will have an understanding of how certain data can be modeled as graphs, and how to apply graph machine learning for effective classification.

Technical requirements

You can find the code files for this chapter on GitHub at `https://github.com/ PacktPublishing/10-Machine-Learning-Blueprints-You-Should-Know-for- Cybersecurity/tree/main/Chapter%208`.

An introduction to graphs

First, let us understand what graphs are and the key terms related to graphs.

What is a graph?

A graph is a data structure that is represented as a set of nodes connected by a set of edges. Mathematically, we specify a graph G as (V, E), where V represents the nodes or vertices and E represents the edges between them, as shown in *Figure 8.1*:

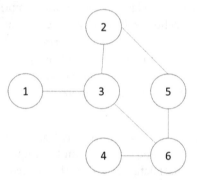

Figure 8.1 – A simple graph

In the previous graph, we have the following:

$V = \{1, 2, 3, 4, 5, 6\}$

$E = \{(1,3), (2,3), (2,5), (3,6), (4,6), (5.6)\}$

Note that the order in which the nodes and edges are mentioned does not matter. The graph shown in *Figure 8.1* is an undirected graph, which means that the direction of the edges does not matter. There can also be directed graphs in which the definition of the edge has some meaning, which gives importance to the direction of the edge. For example, a graph depicting the water flow of from various cities would have directed edges, as water flowing from city A to city B does not imply that water flows from B to A.

An example of a directed graph is shown here:

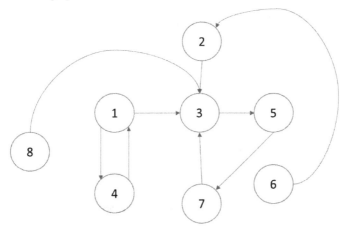

Figure 8.2 – An example of a directed graph

Here, we would define the graph as follows:

$V = \{1, 2, 3, 4, 5, 6, 7, 8\}$

$E = \{(1,4), (4,1), (1,3), (2,3), (3,5), (5,7), (6,2), (7,3), (8,3)\}$

Note that here, as it is a directed graph, the order in which nodes in each edge tuple are specified matters. We have added both $(1,4)$ and $(4,1)$.

In this graph, the edges are unmarked (that is, there is no weight associated with an edge). Such a graph is called an unweighted graph. A graph in which there are weights associated with edges is called a weighted graph. Edge weights can represent properties of relationships between nodes, or the intensity of relationships. For example, in a highway transport network, where nodes represent cities, the edge weights can depict the distance on the highway or the time taken to traverse on average.

Representing graphs

As discussed, a graph can be represented by a set of vertices and edges between vertices. On a computer, or inside a program, a graph can be represented in one of two ways – an adjacency matrix or an adjacency list.

An adjacency matrix is an $N \times N$ matrix, where N is the total number of nodes in the graph. Entries in the matrix denote edge relations between the nodes. If A is the adjacency matrix, then $A_{ij} = 1$ if there is an edge between the nodes i and j. If the graph is undirected, then we have $A_{ij} = A_{ji}$; if the graph is directed, then $A_{ij} = 1$ means that there is an edge from node i to node j

Consider the following directed graph:

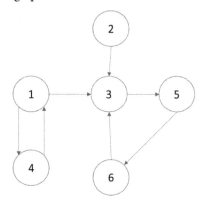

Figure 8.3 – A sample-directed graph

For this directed graph, the adjacency matrix would be defined as follows:

0	0	1	1	0	0
0	0	1	0	0	0
0	0	0	0	1	0
1	0	0	0	0	0
0	0	0	0	0	1
0	0	0	0	0	0

Table 8.1 – An adjacency matrix corresponding to the graph in Figure 8.3

Alternatively, if the graph is weighted, then the entries in the adjacency matrix can be modified to represent the weight instead.

While adjacency matrices are interpretable and straightforward to understand, they are computationally expensive. As the number of nodes in the graph increases, the size of the adjacency matrix also increases. In technical terms, the space complexity is $O(N^2)$, where N is the number of nodes. Larger adjacency matrices are difficult to store and also require a significant processing overhead.

An alternative way to represent a graph is an adjacency list. In this form of representation, the graph is stored as a dictionary or hash table. The keys represent the nodes, and the value is a list of nodes that the key node has an edge to. The list of nodes is represented in memory as a linked list. For the graph in *Figure 8.3*, the adjacency list representation is as follows:

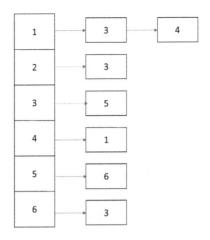

Figure 8.4 – An adjacency list representation of the graph shown in Figure 8.3

Here, we can clearly see the nodes that a given node has an edge to. Even when the number of nodes is large and the edges are sparse, the adjacency list turns out to be an efficient representation.

Graphs in the real world

Much real-world data in nearly all domains of science can be naturally represented as networks using graphs. In the chemical and medical sciences, drugs can be represented as a network of molecules, which are themselves networks of atoms. In information systems and the internet, a graph can be constructed that represents machines and the connections among them. In the field of engineering, a graph can be constructed to represent cities, countries, and the transport networks within them.

Graphs provide us with an aggregate and feature-rich view of data. Whereas, traditionally, every data point would be examined independently, graphs now allow us to examine the relationships between various data points. Relationships and neighborhoods contain valuable information that can be leveraged in recommendations, clustering, fraud detection, and other analytic tasks.

Graphs are rapidly emerging as important analytic tools in the cybersecurity domain. Similar to the other fields, several security applications can be represented as graphs, such as the following:

- A social media network can be represented as a graph with users as nodes. Edges between nodes identify friendships and family relationships.

- Network traffic can be modeled as a graph, with nodes as IP addresses and edges as communication messages.

- Domains, URLs, or websites can be represented as a graph, where edges indicate the presence of links between two websites or domains.

- Malware files can be represented as a graph, where nodes are functions and edges represent calls between them.

Nodes represent entities of interest, and edges represent the relationships between them. Not all nodes have to be of the same category – for example, a social network graph can have both posts and users as nodes. Similarly, there can also be heterogeneity in edges. One kind of edge can connect users who are friends, while another one can connect users who have at least 10 mutual friends. A third one can connect users who have commented on the same post.

Every node in a graph can have features associated with it. These represent characteristics of the entity that the node represents. Features associated with a user can be their age, the age of the account, the number of friends, the number of posts shared, and so on. Similarly, edges can also have features. For example, if an edge represents two users being friends, the edge features can be the number of mutual friends, the age of the friendship, the common pages followed, and so on.

Consider the sample social network graph shown in *Figure 8.5*:

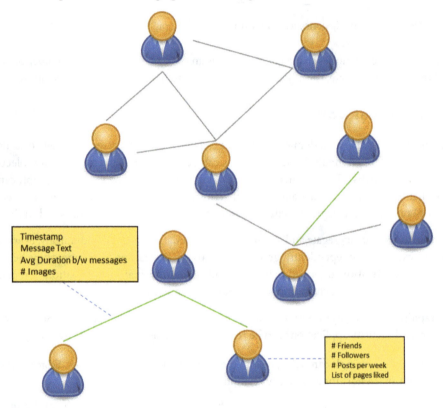

Figure 8.5 – A social network graph

In this graph, the nodes represent users. Every user has features associated with them – the number of friends, followers, and posts per week, as well as a list of pages liked. These become node attributes. There are two kinds of edges, marked by different colors. A green edge is constructed between two users if they have communicated over messages. A gray edge is added if the two have more than five mutual friends. Note that edges can also have features. In this case, the green edge (which denotes the fact that two users talked over messages) has properties related to that relationship, such as message timestamps, the text, the average duration between messages, and the number of images exchanged. These become the edge attributes.

This concludes our discussion on the fundamentals of graphs and how real-world data can naturally be represented in graph form. In the next section, we will look at how machine learning can be applied to graphs.

Machine learning on graphs

Machine learning techniques (such as classification or clustering) can nowadays be applied to nodes, edges, or entire graphs. The concepts remain the same, but we apply the algorithms to graph entities, and therefore, some of the tasks can be framed as a node, link, or subgraph classification. For example, in a network of users on social media, identifying abusive or bot users would be a node classification task. Identifying malicious messages or transactions would be an edge classification problem. Detecting groups of hate speech disseminators would be a graph classification problem.

In graph machine learning, the challenge lies in extracting features from a graph. A possible approach would be using the adjacency matrix and node features as an attribute vector and feeding it to a traditional machine learning algorithm. However, the model produced will not be permutation-invariant, as there is no inherent order within the nodes in a graph; models based on a graph should be permutation-invariant, as the order of nodes should not really matter. There have been several approaches proposed to handle these challenges, which we will take a brief look at.

Traditional graph learning

In a naïve approach to machine learning on graphs, we will parse the graph and extract features for the entity we're interested in. If the task is at the node level, we extract a set of features for each node. If it is at the edge level, we extract a set of features for each edge. However, prior research has shown that this traditional approach is most suited for node-level tasks, such as node classification or clustering. These features are typically based on standard graph-based metrics.

For example, consider a social media network where nodes are users, and edges between users indicate some sort of relationship (e.g., the users are friends, they share a certain number of mutual connections, they have a common or similar activity, or they have interacted via comments or messages).

First, you can extract common graph metrics for every user, such as the following:

- The node in-degree (the number of incoming edges to a node)
- The node out-degree (the number of outgoing edges from a node)
- The sum of outgoing edge weights

- The sum of incoming edge weights

- The distance to the nearest neighbor (as defined by the edge weight)

- The in-degree of the nearest neighbor

- The distance to the root (some predefined neighbor)

- Whether the node is part of a cyclic subgraph (that is, forming a loop)

Additionally, we can use node-specific features that we would have used in traditional machine learning, such as the number of friends, the number of pages liked, the number of logins, the number of posts, the key topics the user writes about, and the average likes per post. Together, these feature sets will form our feature vector.

Once the feature vector has been created, any standard machine learning model (logistic regression, random forest, SVM, or a deep neural network) can be trained for classification.

While this approach is straightforward, it has a major disadvantage – it examines each node individually and does not consider the interrelationships within nodes. Additionally, features have to be handcrafted. Not all graph metrics will make sense in all use cases, so you may have to come up with more creative measurements.

Graph embeddings

The previous section described traditional feature extraction on graphs and the disadvantages that come with it. In this section, we will look at a slightly advanced technique – node embeddings, which are analogous to word embeddings.

Node embeddings

In the previous chapter, we looked at word embeddings using the Word2Vec algorithm. An embedding is a mathematical representation of a word, such that words with a similar meaning or closer in semantics are closer to each other in the embedding space. For example, the words *king* and *queen* will have embeddings that are very similar, as will the words *banana* and *apple*. Node embeddings operate under a similar concept; embeddings are generated for each node such that similar nodes have similar embeddings.

As an example, see the following diagram, taken from the Stanford Course on Graph Neural Networks (CS224W):

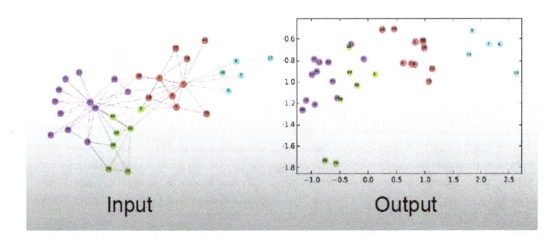

Figure 8.6 – A graph and its corresponding node embeddings

On the left, you can see a graph where nodes are colored, and on the right, you can see the same nodes plotted in an embedding space (a high-dimensional embedding was generated, followed by principal component analysis to reduce it to two dimensions). We can clearly see that nodes of a similar color cluster together in the embedding space.

Node embeddings are generated based on random walks from a node. A random walk is basically the sequence of nodes traversed, starting at a source node. Given a source node, we select a neighbor at random and move to it. At this neighbor, we again select another neighbor (of this new node) and move to it. We do this for a fixed number of steps. The sequence of nodes visited forms a random walk.

We use random walks to generate node embeddings. The underlying rationale is that if a random walk starting at u visits v with a high probability, then the nodes u and v are likely to be similar. Nodes visited in the same random walk will be close to each other in the embedding space.

To generate our embeddings, we estimate the probability of visiting node v on a random walk that started at node u. We learn the mapping from the node to the embedding space. We want to learn feature representations that are predictive of the nodes in the random walk. We first initialize at random an embedding for each node. We then run fixed-length, short random walks and calculate a loss for each walk. The loss function generally used is the log-likelihood. If u is the node of interest, the loss function is defined as follows:

$$L = \sum_{u \in V} \sum_{v \in N_R(u)} -\log(P \mid z_u)$$

Here, z denotes the embedding representation. $N_{R(u)}$ represents the neighborhood of node u. We estimate the probability P of two nodes co-occurring on the same random walk as the softmax of the dot product of their embeddings. The neighborhood set can be formed by any criterion (such as neighbors at one hop, two hops, or satisfying some criterion for similarity).

We optimize for the loss function using gradient descent, just like we would optimize for parameters in any other algorithm, such as a neural network or a linear regression. The embedding z is first initialized to some random value. For each node, a loss is calculated as well as the partial derivative of the loss with respect to z. Finally, z is updated by making adjustments in the right direction as suggested by the gradient. After this has been completed for multiple iterations, z will have the embeddings that result in the smallest possible loss.

The approach described so far was based on random walks. However, a popular algorithm, Node2Vec, operates on the principle of biased walks. Instead of picking the next node to jump to randomly, it operates based on two parameters – p, which denotes the probability of returning to the previous node jumped from, and q, which denotes the probability of moving outward to another node. When the value of p is low, the random walk functions as a **breadth-first search** (**BFS**) (a graph traversal algorithm that explores all the vertices at the same distance from the starting vertex, before moving to the vertices at the next level). On the other hand, when q is low, it functions as a **depth-first search** (**DFS**) (a graph traversal algorithm that explores as far as possible along each branch before backtracking).

From node embeddings to graph embeddings

The random walk and Node2Vec methods describe how to obtain an embedding representation for a single node. However, oftentimes, we want to solve tasks at the graph level, such as classifying subgraphs or entire graphs. There are two approaches that can compute embeddings for graphs.

The first approach is calculating embeddings for each node separately and then aggregating them to obtain a graph-level representation. The aggregation can be as simple as a sum or average. The sum or average can be weighted by node importance, label, or some other feature. While this approach is fairly straightforward, it obfuscates the embeddings of each individual node; it does not take the structure and connections between nodes into account.

Another popular approach is to introduce a dummy node into the graph or subgraph. This node is thought of as having edges to all of the other nodes in the graph. We then use the learned models to calculate the embedding for this dummy node. As this node is connected to all the other nodes, it captures the structural relationships between them and can represent the graph as a whole.

GNNs

Existing methods for machine learning on graphs face the following two issues:

- If we use traditional methods of extracting features based on graph analytics, we lose out on incorporating the node-level features and their relationships. Aggregating metrics at a graph or subgraph level causes information loss, due to noisy data in the model.

- If we use node embeddings, we use only the interconnections between nodes and not the node features. Valuable signals that may have been meaningful for a classification model present in node features will be lost.

This led to the development of GNNs. Using GNN models, it is possible to apply deep learning algorithms (such as convolution, backpropagation, autoencoders, and attention-based transformers) directly to graphs. They accept entire graphs as input (instead of vectors) and learn embedding representations for each node. Every node has an internal state (initially set to a null vector, or based on the features of a node). A node aggregates information from its neighbors, followed by neural message passing, which propagates this information.

Neural message passing is the fundamental principle on which GNNs operate. Consider the following figure, illustrating the operation of message passing and a GNN in a node classification context:

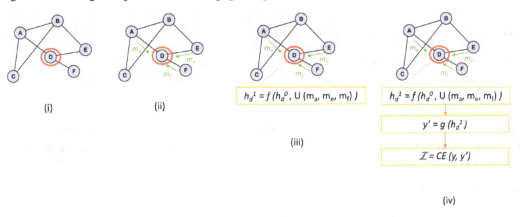

Figure 8.7 – Message passing in GNNs

Consider the graph with nodes and edges, shown in (i). Let **D** be the initial node under consideration. This is an arbitrary choice, and any node can be chosen to start with. Each node has an internal state representation. At first, the internal state is set to the feature vector for the node.

Each neighbor of **D** now passes messages to **D**, as shown in (ii). The messages are simply the internal states, or some function of the internal state. After **D** receives the messages, it does two things, as shown in (iii):

- Aggregates the messages to come up with a unified representation. Aggregation can be a sum, mean, or min/max pooling of the states.

- Updates its own internal state by applying an aggregation function f over the original state and the aggregated state produced in the previous step.

Finally, apply a function g such as a sigmoid or softmax to obtain a prediction for node **D**. Comparing this with the ground truth label for the node, we can calculate a loss, as shown in (iv), which can be backpropagated.

All of these steps ((i)–(iv)) occur for each node at the same instant. The message passing happens first, and the updates of the state occur together. This process can be repeated for multiple iterations.

Note an interesting consequence of message passing. At $t = 0$ (the very first step), a node has information only about itself (based on its own features). At $t = 1$ (after the first step of message passing has completed), a node has received feature information from its neighbors, so it has knowledge about its first neighbors. At $t = 2$, the node will receive messages from its neighbors, but these messages will already encode information from *their* neighbors; thus, a node will have knowledge of its two-hop neighbors as well. As the steps repeat, information spreads farther and farther.

Now that we have described graphs sufficiently, and grasped the concepts behind GNNs, it is time to put them to use.

Fake news detection with GNN

In this section, we will learn how fake news can be detected using a GNN.

Modeling a GNN

While some problems can naturally be thought of as graphs, as data scientists, you need to conceptualize and build a graph. Data may still be available to you in tabular form, but it will be up to you to build a meaningful graph from it.

Solving any task with a GNN involves the following high-level steps:

1. Identifying the entities that will be your nodes.
2. Defining a rule or metric to connect nodes via edges.
3. Defining a set of features for nodes and edges.
4. Determining the kind of graph task the given problem can translate into (node classification, edge classification, or subgraph classification).

In social media-related domains, such as friend recommendation, post virality, and fake news detection, we have multiple choices for nodes, their features, and the methodology for edges between them, such as the following:

- Nodes can be users, their posts, or comments
- Features can be user-historical behavior or text-related features from content
- Edges can be based on whether they have mutual friends or follow the same page

There is a growing body of research that explores these methodologies and applications of GNNs for internet security.

The UPFD framework

For our experiment, we will be following the **User Preference-aware Fake News Detection** (UPFD) framework ([2104.12259] *User Preference-aware Fake News Detection* – `arxiv.org`). In this UPFD framework, every news article is transformed into a graph. The task is to detect whether a news article is fake news or not. As a news article is a graph, this is essentially a graph classification problem.

For every news article, we obtain an information diffusion graph between users. In short, the process works as follows. For every news article, we do the following:

1. Identify the set of users who are engaged in propagating the news article (via likes or retweets). These become the nodes of the graph.

2. Crawl the recent 200 tweets of each user (this is a design choice; you may decide to crawl more or less depending on your use case!).

3. Using pre-trained BERT embeddings and word vectors, encode the news article into a feature vector. This feature vector per user represents the node features.

4. Based on the order of retweeting the news article, build an information diffusion path that indicates how the news article spread from one user to another. These form the edges of the graph.

5. Finally, using the GNN, train a classification model to classify each graph (that is, each article) as fake news or not, based on the graph structure as well as user representations.

The dataset used for UPFD has been made available publicly. This dataset contains graph representations for news articles, after all the pre-processing, feature extraction, and information diffusion has been done. Using this dataset directly saves us the trouble of implementing data preparation pipelines.

Dataset and setup

We will first install the required libraries. As discussed in previous chapters, PyTorch is a deep learning framework developed by Facebook that can help us flexibly and easily implement most machine learning models, including neural networks. PyTorch Geometric is the graph counterpart of PyTorch that can be used to implement GNNs. It is built upon PyTorch and contains several methods for deep learning on graphs. The following commands will install PyTorch and PyTorch Geometric:

```
pip install torch
pip install -q torch-scatter -f https://pytorch-geometric.com/whl/
torch-${TORCH}+${CUDA}.html
pip install -q torch-sparse -f https://pytorch-geometric.com/whl/
torch-${TORCH}+${CUDA}.html
pip install -q git+https://github.com/rusty1s/pytorch_geometric.git
```

We can check that the installation was successful:

```
import torch
import torch_geometric
```

```
version = torch.__version__
print("Torch version is {}".format(version))
```

This will show you the version of PyTorch you are using.

We will now download the fake news dataset from the UPFD paper. This has been integrated with the PyTorch Geometric datasets:

```
from torch_geometric.datasets import UPFD

DATA_ROOT = "/content/FakeNewsNet/dataset"
train_data = UPFD(root = DATA_ROOT,
                  name="gossipcop", feature="content",
                  split="train")
test_data = UPFD(root = DATA_ROOT,
                 name="gossipcop", feature="content",
                 split="test")
```

We can examine the size of our training and test sets:

```
print("# Training Examples: {}".format(len(train_data)))
print("# Test Examples: {}".format(len(test_data)))
```

This should give you the following output:

```
# Training Examples: 1092
# Test Examples: 3826
```

What does the training data look like? Let us look at the first element:

```
train_data[0]
Data(x=[76, 310], edge_index=[2, 75], y=[1])
```

This says that the first element is an object of the Data class. The x attribute represents the features of the nodes.

We can visualize one of the graphs using the networkx library. First, install the library using the package manager:

pip install networkx

Now, we can visualize the graphs using this library:

```
nx.draw(to_networkx(train_data[1]))
```

It will produce the following result:

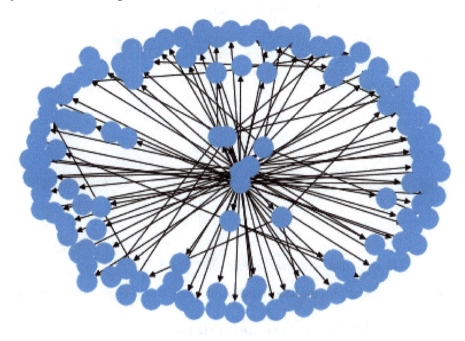

Figure 8.8 – A sample graph from the UPFD dataset

On some systems, or if you are using Google Colab, the `networkx` dependencies do not load properly, and you might get an error when visualizing the graph that says that the function is not defined. If that happens, you need to manually copy the code snippet of that function. For convenience, the function definition is given here:

```
def to_networkx(data, node_attrs=None, edge_attrs=None, to_
undirected=False,
              remove_self_loops=False):
    if to_undirected:
        G = nx.Graph()
    else:
        G = nx.DiGraph()
    G.add_nodes_from(range(data.num_nodes))
    node_attrs, edge_attrs = node_attrs or [], edge_attrs or []
    values = {}
    for key, item in data(*(node_attrs + edge_attrs)):
        if torch.is_tensor(item):
            values[key] = item.squeeze().tolist()
        else:
```

```
                values[key] = item
          if isinstance(values[key], (list, tuple)) and len(values[key])
== 1:
                values[key] = item[0]
       for i, (u, v) in enumerate(data.edge_index.t().tolist()):
          if to_undirected and v > u:
              continue
          if remove_self_loops and u == v:
              continue
          G.add_edge(u, v)
          for key in edge_attrs:
              G[u][v][key] = values[key][i]
      for key in node_attrs:
          for i, feat_dict in G.nodes(data=True):
              feat_dict.update({key: values[key][i]})
      return G
```

Copying this into your script should resolve the error.

Now that we are familiar with what the data looks like internally, we can begin our experiment.

Implementing GNN-based fake news detection

First, we read the data into the DataLoader object so that it can easily be consumed by our model. The batch_size parameter depicts the number of examples that will be processed in the same batch:

```
from torch_geometric.loader import DataLoader
train_loader = DataLoader(train_data, batch_size=128, shuffle=True)
test_loader = DataLoader(test_data, batch_size=128, shuffle=False)
```

Now, we will define the actual GNN:

```
from torch_geometric.nn import global_max_pool as gmp
from torch_geometric.nn import GATConv
from torch.nn import Linear

class GNN(torch.nn.Module):
    def __init__(self,
                 n_in, n_hidden, n_out):

        super().__init__()

        # Graph Convolutions
        self.convolution_1 = GATConv(n_in, n_hidden)
        self.convolution_2 = GATConv(n_hidden, n_hidden)
```

```
        self.convolution_3 = GATConv(n_hidden, n_hidden)

        # Readout Layers
        # For news features
        self.lin_news = Linear(n_in, n_hidden)

        # For processing graph features
        self.lin0 = Linear(n_hidden, n_hidden)

        # For pre-final layer for softmax
        self.lin1 = Linear(2*n_hidden, n_out)

    def forward(self, x, edge_index, batch):
        # Graph Convolutions
        h = self.conv1(x, edge_index).relu()
        h = self.conv2(h, edge_index).relu()
        h = self.conv3(h, edge_index).relu()

        # Pooling
        h = gmp(h, batch)

        # Readout
        h = self.lin0(h).relu()

        # Following the UPFD paper, we include raw word2vec embeddings
of news
        root = (batch[1:] - batch[:-1]).nonzero(as_tuple=False).view(-
1)
        root = torch.cat([root.new_zeros(1), root + 1], dim=0)

        news = x[root]
        news = self.lin_news(news).relu()

        out = self.lin1(torch.cat([h, news], dim=-1))
        return torch.sigmoid(out)
```

Let us deconstruct this code a little bit. At a high level, we write a class to define our GNN. The class has two functions – a constructor that creates objects and sets initial parameters (__init__), and a method that defines the structure of the network by laying out, step by step, the transformations that happen in the forward pass of the neural network.

The constructor function takes in three parameters. These are hyperparameters and are chosen for optimal performance. The first parameter is the number of input features. The second parameter defines the size of the hidden layer. And the final parameter defines the number of neurons in the output layer (typically, either 1, in the case of binary classification, or equal to the number of classes in multi-class classification).

The constructor defines the layers of the neural network we will need to process our data. First, we define three convolutional layers. Then, we define our readout layers. We will need three separate readout layers:

- One for the news features
- One for the graph features
- One to obtain softmax probability distributions

The `forward()` function defines the steps taken in the forward pass of the neural network. In other words, it describes how the input features are transformed into the output label.

First, we see that the features are processed through the first convolution layer. The output is passed to a second convolution, and the output of this is passed to a third. Then, a pooling layer aggregates the output of this final convolution. The first readout layer will process the pooled output. This completes the processing of graph features.

We will follow the same philosophy outlined in the UPFD paper. We will extract news features as well. Going by the design of the dataset, the first node is the root node in the graph. We extract features from this node and pass them through the readout layer we defined for the news features.

Finally, we concatenate the two – the readout output of the graph features and the readout output of the news features. This concatenated vector is passed through a final readout layer. We transform the output of this using a sigmoid function (as it is a binary classification) to obtain the output class label probability.

Now that we have defined the class and the structure of the GNN, we can actually train the model! We begin by defining some important terms we need:

```
if torch.cuda.is_available():
  device = 'cuda'
else:
  device = 'cpu'

model = GNN(train_data.num_features, 128, 1).to(device)
optimizer = torch.optim.Adam(model.parameters(),
                             lr=0.01, weight_decay=0.01)
loss_fnc = torch.nn.BCELoss()
```

First, we check whether a GPU is available via CUDA. If it is, we set `'cuda'` as the device, as opposed to the usual `'cpu'`. Using CUDA or any GPU significantly speeds up the model training, as matrix multiplication, gradient calculation, and backpropagation can be decoupled and parallelized.

Then, we define three key objects we need:

- **Model**: We defined the GNN class that describes the model structure and operations in the forward pass. We use this class to create a GNN object.

- **Optimizer**: We will calculate the output in the forward pass, compute a loss, and backpropagate with the gradients. The optimizer will help us perform gradient descent to minimize loss.

- **Loss**: The core of a neural network (or any machine learning model) is to minimize a loss so that the predicted values are as close to the actual values as possible. Here, for the loss function, we will use the binary cross-entropy loss, as it is a binary classification problem.

Now, we write the actual training function:

```
def train(epoch):
    model.train()
    total_loss = 0
    for data in train_loader:
        data = data.to(device)
        optimizer.zero_grad()
        out = model(data.x, data.edge_index, data.batch)
        loss = loss_fnc(torch.reshape(out, (-1,)), data.y.float())
        loss.backward()
        optimizer.step()
        total_loss += float(loss)    data.num_graphs
    return total_loss / len(train_loader.dataset)
```

Let us deconstruct what we have written in the training function. Recall that the basic steps in a neural network are as follows:

1. Pass the features input to the neural network.

2. Process the features through the various layers till you reach the output.

3. Once the output is derived, compute the loss by comparing it with the ground truth label.

4. Backpropagate the loss and adjust each of the parameters in the layers using gradient descent.

5. Repeat *steps 1–4* for each example to complete one epoch.

6. Repeat *steps 1–5* for multiple epochs to minimize the loss.

The previous training code merely replicates these steps. We read the input features and copy them over to the GPU if needed. When we run them through the model, we basically do the forward pass. Based on the output, we calculate the loss and then initiate the backward propagation.

We will also define a function that calculates metrics for us. Recall that we have previously used the confusion matrix to compute certain metrics, such as accuracy, precision, recall, and the F-1 score. In this example, we will use the built-in modules provided by scikit-learn. These functions will take in the list of predicted and actual values, and compare them to calculate a given metric.

Our metric function is as follows:

```
from sklearn.metrics import accuracy_score, f1_score
def metrics(predicted, actuals):
    preds = torch.round(torch.cat(predicted))
    acts = torch.cat(actuals)
    acc = accuracy_score(preds, acts)
    f1 = f1_score(preds, acts)
    return acc, f1
```

We have already written a function that will train the model. We will now write a similar function but just for inference. We first set the model in the evaluation mode using the `eval()` function. In the evaluation mode, there is no gradient calculation (as there is no training involved, we do not need the gradients for backpropagation; this saves computation time as well as memory).

Similar to the training function, the evaluation function reads a batch of data and moves it to the GPU if needed and available. It runs the data through the model and calculates the predicted output class. The predicted labels from the model output and actual labels from the ground truth are appended to two separate lists. Based on the model output and the ground truth, we calculate the loss. We repeat this for each batch and sum up the loss values.

At the end, when all the data has been processed, we have the average loss (the total loss divided by the number of data points). We use the actual and predicted lists to calculate the metrics using the metric function defined earlier:

```
@torch.no_grad()
def test(epoch):
    model.eval()
    total_loss = 0
    all_preds = []
    all_labels = []
    for data in test_loader:
        data = data.to(device)
        out = model(data.x, data.edge_index, data.batch)
        loss = loss_fnc(torch.reshape(out, (-1,)), data.y.float())
        total_loss += float(loss)   data.num_graphs
        all_preds.append(torch.reshape(out, (-1,)))
        all_labels.append(data.y.float())
```

```
    # Calculate Metrics
    accuracy, f1 = metrics(all_preds, all_labels)
    avg_loss = total_loss/len(test_loader.dataset)

    return avg_loss, accuracy, f1
```

Finally, it is time to put all these functions into action! So far, we have the following:

- A class that defines our GNN structure

- A function that will run a training iteration over the GNN model

- A function that will run an evaluation/inference iteration on the model

- A function that will compute common metrics

We will now use all of these to run our GNN experiment. We will first run the training function, and obtain a loss. We will run the test function and obtain the test loss. We repeat this for multiple epochs. As the epochs progress, we can observe how the loss changes in both the training and test data:

```
NUM_EPOCHS = 50
train_losses = []
test_losses = []
for epoch in range(NUM_EPOCHS):
    train_loss = train(epoch)
    test_loss, test_acc, test_f1 = test(epoch)
    train_losses.append(train_loss)
    test_losses.append(test_loss)
    print(f'Epoch: {epoch:04d}  ==  Training Loss: {train_
loss:.4f}  ==  '
          f'TestLoss: {test_loss:.4f}  ==  TestAcc: {test_
acc:.4f}  ==  TestF1: {test_f1:.4f}')
```

This should produce something like this:

```
Epoch: 0000  ==  Training Loss: 0.7314  ==  TestLoss: 0.6950  ==  TestAcc: 0.4992  ==  TestF1: 0.0000
Epoch: 0001  ==  Training Loss: 0.6998  ==  TestLoss: 0.6994  ==  TestAcc: 0.4992  ==  TestF1: 0.0000
Epoch: 0002  ==  Training Loss: 0.6936  ==  TestLoss: 0.6962  ==  TestAcc: 0.5008  ==  TestF1: 0.6674
Epoch: 0003  ==  Training Loss: 0.6969  ==  TestLoss: 0.7393  ==  TestAcc: 0.5008  ==  TestF1: 0.6674
Epoch: 0004  ==  Training Loss: 0.7103  ==  TestLoss: 0.6903  ==  TestAcc: 0.4992  ==  TestF1: 0.0000
Epoch: 0005  ==  Training Loss: 0.6995  ==  TestLoss: 0.6849  ==  TestAcc: 0.9127  ==  TestF1: 0.9144
Epoch: 0006  ==  Training Loss: 0.6881  ==  TestLoss: 0.6981  ==  TestAcc: 0.4992  ==  TestF1: 0.0000
Epoch: 0007  ==  Training Loss: 0.7061  ==  TestLoss: 0.6876  ==  TestAcc: 0.4992  ==  TestF1: 0.0000
Epoch: 0008  ==  Training Loss: 0.6846  ==  TestLoss: 0.7290  ==  TestAcc: 0.4992  ==  TestF1: 0.0000
Epoch: 0009  ==  Training Loss: 0.6942  ==  TestLoss: 0.7698  ==  TestAcc: 0.4992  ==  TestF1: 0.0000
Epoch: 0010  ==  Training Loss: 0.7045  ==  TestLoss: 0.7112  ==  TestAcc: 0.5008  ==  TestF1: 0.6674
Epoch: 0011  ==  Training Loss: 0.7214  ==  TestLoss: 0.7261  ==  TestAcc: 0.4992  ==  TestF1: 0.0000
Epoch: 0012  ==  Training Loss: 0.6905  ==  TestLoss: 0.6918  ==  TestAcc: 0.5008  ==  TestF1: 0.6674
Epoch: 0013  ==  Training Loss: 0.6891  ==  TestLoss: 0.6817  ==  TestAcc: 0.4992  ==  TestF1: 0.0000
Epoch: 0014  ==  Training Loss: 0.6907  ==  TestLoss: 0.6892  ==  TestAcc: 0.4992  ==  TestF1: 0.0000
Epoch: 0015  ==  Training Loss: 0.6815  ==  TestLoss: 0.6751  ==  TestAcc: 0.5008  ==  TestF1: 0.6674
Epoch: 0016  ==  Training Loss: 0.6769  ==  TestLoss: 0.6698  ==  TestAcc: 0.8006  ==  TestF1: 0.7566
Epoch: 0017  ==  Training Loss: 0.6720  ==  TestLoss: 0.6732  ==  TestAcc: 0.5008  ==  TestF1: 0.6674
Epoch: 0018  ==  Training Loss: 0.6950  ==  TestLoss: 0.6770  ==  TestAcc: 0.4992  ==  TestF1: 0.0000
Epoch: 0019  ==  Training Loss: 0.6624  ==  TestLoss: 0.6567  ==  TestAcc: 0.6189  ==  TestF1: 0.7240
Epoch: 0020  ==  Training Loss: 0.6680  ==  TestLoss: 0.6999  ==  TestAcc: 0.4992  ==  TestF1: 0.0000
Epoch: 0021  ==  Training Loss: 0.6509  ==  TestLoss: 0.6517  ==  TestAcc: 0.5008  ==  TestF1: 0.6674
Epoch: 0022  ==  Training Loss: 0.6385  ==  TestLoss: 0.6239  ==  TestAcc: 0.8547  ==  TestF1: 0.8712
Epoch: 0023  ==  Training Loss: 0.6256  ==  TestLoss: 0.6718  ==  TestAcc: 0.5021  ==  TestF1: 0.0114
Epoch: 0024  ==  Training Loss: 0.6490  ==  TestLoss: 0.6415  ==  TestAcc: 0.5008  ==  TestF1: 0.6674
Epoch: 0025  ==  Training Loss: 0.6138  ==  TestLoss: 0.5853  ==  TestAcc: 0.7261  ==  TestF1: 0.6270
Epoch: 0026  ==  Training Loss: 0.5742  ==  TestLoss: 0.5619  ==  TestAcc: 0.8722  ==  TestF1: 0.8596
Epoch: 0027  ==  Training Loss: 0.5778  ==  TestLoss: 0.5604  ==  TestAcc: 0.6892  ==  TestF1: 0.5525
Epoch: 0028  ==  Training Loss: 0.6164  ==  TestLoss: 0.5497  ==  TestAcc: 0.7020  ==  TestF1: 0.5793
Epoch: 0029  ==  Training Loss: 0.5534  ==  TestLoss: 0.5429  ==  TestAcc: 0.7047  ==  TestF1: 0.5846
Epoch: 0030  ==  Training Loss: 0.5218  ==  TestLoss: 0.5898  ==  TestAcc: 0.5826  ==  TestF1: 0.2855
Epoch: 0031  ==  Training Loss: 0.5204  ==  TestLoss: 0.4884  ==  TestAcc: 0.9177  ==  TestF1: 0.9208
Epoch: 0032  ==  Training Loss: 0.5045  ==  TestLoss: 0.5252  ==  TestAcc: 0.6688  ==  TestF1: 0.5080
Epoch: 0033  ==  Training Loss: 0.5065  ==  TestLoss: 0.4625  ==  TestAcc: 0.9036  ==  TestF1: 0.9100
Epoch: 0034  ==  Training Loss: 0.5405  ==  TestLoss: 0.5971  ==  TestAcc: 0.5844  ==  TestF1: 0.2908
Epoch: 0035  ==  Training Loss: 0.5071  ==  TestLoss: 0.4913  ==  TestAcc: 0.7125  ==  TestF1: 0.6000
Epoch: 0036  ==  Training Loss: 0.4475  ==  TestLoss: 0.4372  ==  TestAcc: 0.9062  ==  TestF1: 0.9123
Epoch: 0037  ==  Training Loss: 0.4357  ==  TestLoss: 0.4363  ==  TestAcc: 0.8758  ==  TestF1: 0.8879
Epoch: 0038  ==  Training Loss: 0.4310  ==  TestLoss: 0.5957  ==  TestAcc: 0.5233  ==  TestF1: 0.6775
Epoch: 0039  ==  Training Loss: 0.4664  ==  TestLoss: 0.4543  ==  TestAcc: 0.7961  ==  TestF1: 0.8292
Epoch: 0040  ==  Training Loss: 0.4424  ==  TestLoss: 0.4032  ==  TestAcc: 0.8975  ==  TestF1: 0.9054
Epoch: 0041  ==  Training Loss: 0.4115  ==  TestLoss: 0.3810  ==  TestAcc: 0.9276  ==  TestF1: 0.9297
Epoch: 0042  ==  Training Loss: 0.3706  ==  TestLoss: 0.4598  ==  TestAcc: 0.7627  ==  TestF1: 0.8072
Epoch: 0043  ==  Training Loss: 0.4089  ==  TestLoss: 0.3916  ==  TestAcc: 0.8777  ==  TestF1: 0.8895
Epoch: 0044  ==  Training Loss: 0.3668  ==  TestLoss: 0.4023  ==  TestAcc: 0.8053  ==  TestF1: 0.7621
Epoch: 0045  ==  Training Loss: 0.3820  ==  TestLoss: 0.3446  ==  TestAcc: 0.9318  ==  TestF1: 0.9323
Epoch: 0046  ==  Training Loss: 0.3601  ==  TestLoss: 0.3455  ==  TestAcc: 0.9252  ==  TestF1: 0.9286
Epoch: 0047  ==  Training Loss: 0.3435  ==  TestLoss: 0.3410  ==  TestAcc: 0.9234  ==  TestF1: 0.9272
Epoch: 0048  ==  Training Loss: 0.3394  ==  TestLoss: 0.3476  ==  TestAcc: 0.9064  ==  TestF1: 0.9129
Epoch: 0049  ==  Training Loss: 0.3220  ==  TestLoss: 0.3191  ==  TestAcc: 0.9336  ==  TestF1: 0.9346
```

Figure 8.9 – The training loop of the GNN

We can manually observe the loss trends. However, it would be easier if we could visualize it as a time series over epochs:

```
import matplotlib.pyplot as plt
plt.plot(list(range(NUM_EPOCHS)),
        train_losses,
        color = 'blue', label = 'Training Loss')
plt.plot(list(range(NUM_EPOCHS)),
        test_losses,
        color = 'red', label = 'Test Loss')
plt.xlabel("Epoch")
plt.ylabel("Loss")
plt.legend()
```

You will get the following result:

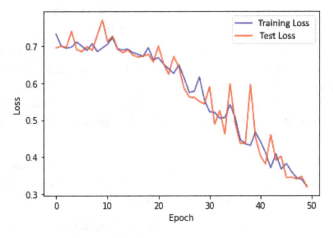

Figure 8.10 – Loss trends across epochs

We see that as the epochs progressed, the loss steadily decreased for the training data. The test loss has also decreased, but there are some spikes, probably caused by overfitting.

Playing around with the model

As in the previous chapters, this model also makes several design choices that affect its performance. We will leave it up to you to try out some changes and observe model performance. Some things to be considered are as follows:

- **Hyperparameters**: The learning rate in the training function and the number of epochs are simply design choices. What happens if they are changed? How do the training and test loss trends change? How do the accuracy and F-1 score change?

- **Graph layers**: The previous example uses the **GATConv** layers; however, we can just as easily use other types of layers available in the PyTorch Geometric library. Read up on these layers – which other layers are suitable? How does using a different layer affect the performance?

- **Architecture**: We used three convolutional layers and three graph layers in our model. However, we can go as high as we want. What happens if we use five layers? Ten layers? Twenty layers? What if we increase only the convolution layers and not the graph layers, or vice versa?

This completes our discussion of how to model data as graphs, and how GNNs can be used to detect fake news.

Summary

A lot of real-world data can be naturally represented as graphs. Graphs are especially important in a social network context where multiple entities (users, posts, or media) are linked together, forming natural graphs. In recent times, the spread of misinformation and fake news is a problem of growing concern. This chapter focused on detecting fake news using GNNs.

We began by first learning some basic concepts about graphs and techniques to learn on graphs. This included using static features extracted from graph analytics (such as degrees and path lengths), node and graph embeddings, and finally, neural message passing, using GNNs. We looked at the UPFD framework and how a graph can be built for a news article, complete with node features that incorporate historical user behavior. Finally, we trained a GNN model to build a graph classifier that detects whether a news article is fake or not.

In the field of cybersecurity, graphs are especially important. This is because attacks are often coordinated from different entities, or directed toward a set of similar targets. Designing a graph and implementing a GNN-based classifier is a valuable skill for data scientists in the security domain.

In the next chapter, we will look at adversarial machine learning and how it can be used to fool or attack machine learning models.

9

Attacking Models with Adversarial Machine Learning

Recent advances in **machine learning** (**ML**) and **artificial intelligence** (**AI**) have increased our reliance on intelligent algorithms and systems. ML systems are used to make decisions on the fly in several critical applications. For example, whether a credit card transaction should be authorized or not or whether a particular Twitter account is a bot or not is decided by a model within seconds, and this decision affects steps taken in the real world (such as the transaction or account being flagged as fraudulent). Attackers use the reduced human involvement to their advantage and aim to attack models deployed in the real world. **Adversarial ML** (**AML**) is a field of ML that focuses on detecting and exploiting flaws in ML models.

Adversarial attacks can come in several forms. Attackers may try to manipulate the features of a data point so that it is misclassified by the model. Another threat vector is data poisoning, where attackers introduce perturbations into the training data itself so that the model learns from incorrect data and thus performs poorly. An attacker may also attempt to run membership inference attacks to determine whether an individual was included in the training data or not. Protecting ML models from adversarial attacks and, therefore, understanding the nature and workings of such attacks is essential for data scientists in the cybersecurity space.

In this chapter, we will cover the following main topics:

- Introduction to AML
- Attacking image models
- Attacking text models
- Developing robustness against adversarial attacks

This chapter will help you understand how adversarial attacks can manifest themselves, which will then help you uncover gaps and vulnerabilities in your ML infrastructure.

Technical requirements

You can find the code files for this chapter on GitHub at `https://github.com/ PacktPublishing/10-Machine-Learning-Blueprints-You-Should-Know-for- Cybersecurity/tree/main/Chapter%209`.

Introduction to AML

In this section, we will learn about what AML exactly is. We will begin by understanding the importance ML plays in today's world, followed by the various kinds of adversarial attacks on models.

The importance of ML

In recent times, our reliance on ML has increased. Automated systems and models are in every sphere of our life. These systems often allow for fast decision-making without the need for manual human intervention. ML is a boon to security tasks; a model can learn from historical behavior, identify and recognize patterns, extract features, and render a decision much faster and more efficiently than a human can. Examples of some ML systems handling security-critical decisions are given here:

- Real-time fraud detection in credit card usage often uses ML. Whenever a transaction is made, the model looks at your location, the amount, the billing code, your past transactions, historical patterns, and other behavioral features. These are fed into an ML model, which will render a decision of *FRAUD* or *NOT FRAUD*.

- Malware detection systems use ML models to detect malicious applications. The model uses API calls made, permissions requested, domains connected, and so on to classify the application as *MALWARE* or *BENIGN*.

- Social media platforms use ML to identify hate speech or toxic content. Models can extract text and image content, topics, keywords, and URLs to determine whether a post is *TOXIC* or *NON-TOXIC*.

What is the goal behind listing these applications? In each case, you can see that ML plays a prominent role in detecting or identifying an adversary or attacker. The attacker, therefore, has an incentive to degrade the performance of the model. This leads us to the branch of AML and adversarial attacks.

Adversarial attacks

AML is a subfield of AI and ML concerned with the design and analysis of algorithms that can robustly and securely operate in adversarial environments. In these scenarios, an adversary with malicious intent can manipulate input data to disrupt the behavior of ML models, either by causing incorrect predictions or by compromising the confidentiality and privacy of the data.

AML attacks are intentionally crafted inputs to ML models that cause them to behave in unintended and potentially harmful ways. They can be used for malicious purposes, such as compromising the security and privacy of ML models, disrupting their normal operation, or undermining their accuracy and reliability. In these scenarios, attackers may use adversarial examples to trick ML models into making incorrect predictions, compromising the confidentiality and privacy of sensitive data, or causing harm to the system or the people who use it.

For example, we listed the applications of ML models in some critical tasks earlier. Here is how an attacker could manipulate them to their benefit:

- In a fraud detection system, a smart attacker may try to abuse the credit card with multiple small purchases instead of a large one. The model may be fooled by the purchase amounts and will not flag them as abnormal. Or, the attacker may use a **virtual private network** (**VPN**) connection to appear closer to the victim and purchase gift cards online, thus evading the model's location-based features.

- A malware developer may know which features indicate malware presence. Therefore, they may try to mask that behavior by requesting some normal permissions or making redundant API calls so as to throw the classifier off in the predictions.

- A user who wants to post toxic content or hate speech knows which words indicate abusive content. They will try to misspell those words so that they are not flagged by the model.

Using adversarial attacks, an attacker can potentially fool a system and escape undetected. It is, therefore, important for researchers and practitioners in the field of ML to understand adversarial attacks and to develop methods for detecting and defending against them. This requires a deep understanding of the underlying mechanisms of these attacks and the development of new algorithms and techniques to prevent them.

Adversarial tactics

The end goal of an adversarial attack is to degrade the performance of an ML model. Adversarial attacks generally employ one of three strategies: input perturbation, data poisoning, or model inversion attacks. We will cover these in detail next.

Input perturbation attacks

In input perturbation attacks, the attacker maliciously crafts input examples so that they will be misclassified by the model. The attacker makes slight changes to the input that are neither discernible to the naked eye nor large enough to be detected as anomalous or noisy. Typically, this can include changing a few pixels in an image or altering some characters in a word. **Deep learning** (**DL**) systems, favored because of their power, are very susceptible to input perturbation attacks. Because of non-linear functions and transforms, a small change in the input can cause a significant unexpected change in the output.

For example, consider the following screenshot from a 2017 study showing two images of a **STOP** sign:

Figure 9.1 – An actual image of a STOP sign (left) and the adversarially manipulated image (right)

The one on the left is the actual one from the street, and the one on the right is an identical one with some pieces of tape. These pieces of tape represent an input perturbation on the original. The researchers found that the image on the left was correctly detected as a STOP sign, but the model was fooled by the right one and detected it as a 45 MPH speed limit sign.

Data poisoning attacks

Data poisoning attacks are malicious attacks in which an adversary manipulates or corrupts training data in order to degrade the performance of an ML model or cause it to behave in unexpected ways. The goal of these attacks is to cause the model to make incorrect predictions or decisions, leading to security vulnerabilities or privacy breaches. If the quality of data (in terms of the correctness of labels presented to the model) is bad, naturally the resulting model will also be bad. Due to incorrect labels, the model will learn correlations and features incorrectly.

For example, in a **supervised ML (SML)** scenario, an adversary may manipulate the labeled data used for training in order to cause a classifier to misclassify certain instances, leading to security vulnerabilities. In another scenario, an adversary may add malicious instances to the training data in order to cause the model to overfit, leading to a decrease in performance on unseen data. For example, if an attacker adds several requests from a malicious domain and marks them as safe or benign in the training data, the model may learn that this domain indicates safe behavior and, therefore, will not mark other requests from that domain as malicious.

These attacks can be particularly dangerous because ML models are becoming increasingly widely used in a variety of domains, including security and privacy-sensitive applications.

Model inversion attacks

Model inversion attacks are a type of privacy attack in which an adversary tries to reverse-engineer an ML model to obtain sensitive information about the training data or the individuals represented by the data. The goal of these attacks is to reveal information about the training data that would otherwise be protected.

For example, in a fraud detection scenario, a financial institution might use an ML model to identify instances of fraud in financial transactions. An adversary might attempt to reverse-engineer the model in order to obtain information about the characteristics of fraudulent transactions, such as the types of goods or services that are commonly purchased in fraud cases. The attacker may discover the important features that are used to discern fraud and, therefore, know what to manipulate. This information could then be used to commit more sophisticated fraud in the future.

To carry out a model inversion attack, an adversary might start by submitting queries to the ML model with various inputs and observing the model's predictions. Over time, the adversary could use this information to build an approximation of the model's internal representation of the data. In some cases, the adversary might be able to obtain information about the training data itself, such as the values of sensitive features—for example, the age or address of individuals represented by the data.

This concludes our discussion of various kinds of adversarial attacks. In the next section, we will turn to implementing a few adversarial attacks on image-based models.

Attacking image models

In this section, we will look at two popular attacks on image classification systems: **Fast Gradient Sign Method** (**FGSM**) and the **Projected Gradient Descent** (**PGD**) method. We will first look at the theoretical concepts underlying each attack, followed by actual implementation in Python.

FGSM

FGSM is one of the earliest methods used for crafting adversarial examples for image classification models. Proposed by Goodfellow in 2014, it is a simple and powerful attack against **neural network** (**NN**)-based image classifiers.

FGSM working

Recall that NNs are layers of neurons placed one after the other, and there are connections from neurons in one layer to the next. Each connection has an associated weight, and the weights represent the model parameters. The final layer produces an output that can be compared with the available ground truth to calculate the loss, which is a measure of how far off the prediction is from the actual ground truth. The loss is *backpropagated*, and the model *learns* by adjusting the parameters based on the gradient of the loss. This process is known as *gradient descent*. If θ is the parameter and L is the loss, the adjusted parameter θ' is calculated as follows:

$$\theta' = \theta - \eta \frac{\delta L}{\delta \theta}$$

Here, the derivative term $\frac{\delta L}{\delta \theta}$ is known as the *gradient*.

The FGSM adversarial attack leverages the gradients to craft adversarial examples. While the learning algorithm *minimizes* the loss by adjusting the weights, the FGSM attack works to adjust the input data so as to *maximize* the loss. During backpropagation, a small perturbation is added to the image based on the sign of the gradient.

Formally speaking, given an image X, a new (adversarial) image X' can be calculated as follows:

$$X' = X + \epsilon \cdot sign\left(\nabla_x \mathscr{L}(\theta, x, y)\right)$$

Here, $\mathscr{L}$ represents the loss, θ represents the model parameters, and y refers to the ground-truth label. The term $\mathscr{L}(\theta, x, y)$ calculates the loss based on the model prediction and ground truth, and ∇_x calculates the gradient. The term ϵ is the perturbation added, which is either positive or negative depending on the sign of the gradient.

A popular example that demonstrates the effectiveness of the FGSM attack is shown here:

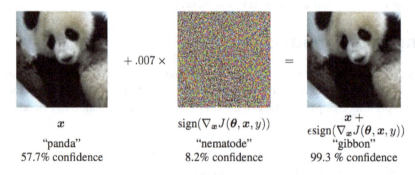

x
"panda"
57.7% confidence

$sign(\nabla_x J(\theta, x, y))$
"nematode"
8.2% confidence

$x + \epsilon sign(\nabla_x J(\theta, x, y))$
"gibbon"
99.3 % confidence

Figure 9.2 – Adding noise to an image with FGSM

The original image is predicted to be a panda with a confidence of 57.7%, which indicates that the model made the correct prediction. Adversarial noise is added to the image depending on the sign of the gradient per pixel (so, either 0.007, 0, or -0.007). The resulting image is identical to the original one—no difference can be seen to the naked eye. This is expected because the human eye is not sensitive to such small differences at the pixel level. However, the model now predicts the image to be a gibbon with a 99.3% confidence.

FGSM implementation

Let us now implement the FGSM attack method using the PyTorch library.

We begin, as usual, by importing the required libraries:

```
import torch
import torch.nn as nn
import torch.optim as optim
import numpy as np
```

```
import torch.nn.functional as F
import matplotlib.pyplot as plt
from torchvision import datasets, transforms
```

Next, we will define a function that performs the FGSM attack and generates an adversarial example. This function calculates the gradient, followed by the perturbation, and generates an adversarial image. The epsilon parameter is passed to it, which indicates the degree of perturbation to be added. In short, the function works as follows:

1. Takes in as input an image or an array of images.

2. Calculates the predicted label by running it through the model.

3. Calculates the loss by comparing the predicted label with the actual ground truth.

4. Backpropagates the loss and calculates the gradient.

5. Calculates the perturbation by multiplying epsilon with the sign of the gradient.

6. Adds this to the image to obtain the adversarial image.

The following code snippet shows how to perform the FGSM attack:

```
def Generate_FGSM_Image(model,
                        x,
                        epsilon):

  # Check if epsilon is 0
  # If so, that means no perturbation is added
  # We can avoid gradient calculations
  if epsilon == 0:
    return x

  # Convert x to a float and having gradients enabled
  x = x.clone().detach()
  x = x.to(torch.float)
  x - x.requires_grad_(True)

  # Get original label as predicted by model
  _, y = torch.max(model(x), 1)

  # Compute Loss
  loss_function = nn.CrossEntropyLoss()
  loss = loss_function(model(x), y)

  # Backpropagate Loss
  loss.backward()
```

```
# Calculate perturbation using the FGSM equation
perturbation = epsilon * torch.sign(x.grad)

# Calculate the adversarial image
x_adversarial = x + perturbation

return x_adversarial
```

Next, we need a basic image classifier to use as the model to attack. As the data at hand is images, we will use a **convolutional neural network** (**CNN**). For this, we define a class that has two functions, as follows:

- **The constructor**: This function defines the basic structure of the NN to be used for the classification. We define the convolutional and fully connected layers that we need. The number of neurons and the number of layers here are all design choices.

- **The forward function**: This function defines what happens during the forward pass of the NN. We take in the input data and pass it through the first convolutional layer. The output of this layer is processed through a ReLU activation function and then passed to the next layer. This continues for all convolutional layers we have. Finally, we flatten the output of the last convolutional layer and pass it through the fully connected layer.

The process is illustrated in the following code snippet:

```
class BasicImageNetCNN(nn.Module):

    def __init__(self, in_channels=1):
        super(BasicImageNetCNN, self).__init__()

        # Define the convolutional layers
        self.conv1 = nn.Conv2d(in_channels, 64, 8, 1)
        self.conv2 = nn.Conv2d(64, 128, 6, 2)
        self.conv3 = nn.Conv2d(128, 128, 5, 2)

        # Define the fully connected layer
        self.fc = nn.Linear(128 * 3 * 3, 10)

    def forward(self, x):

        # Pass the image through convolutional layers one by one
        x = F.relu(self.conv1(x))
        x = F.relu(self.conv2(x))
        x = F.relu(self.conv3(x))
```

```
            # Flatten the output of the convolutional layer and pass to
fully connected layer
            x = x.view(-1, 128 * 3 * 3)
            x = self.fc(x)


            return x
```

We will now write a function that loads the required datasets. For our experiment, we will be using the *CIFAR-10* dataset. Developed by the **Canadian Institute for Advanced Research** (**CIFAR**), the dataset consists of 60,000 images from 10 different classes (airplane, automobile, bird, cat, deer, dog, frog, horse, ship, and truck). Each image is color and of size 32 x 32 pixels. The dataset has been divided into 50,000 training and 10,000 test images. As a standardized dataset in the world of ML, it is well integrated with Python and PyTorch. The following function provides the code to load the train and test sets. If the data is not locally available, it will first download it and then load it:

```
def load_cifar10_datasets(datapath):

    # Load the transformations
    train_transforms = torchvision.transforms.Compose([torchvision.
transforms.ToTensor()])
    test_transforms = torchvision.transforms.Compose([torchvision.
transforms.ToTensor()])

    # Obtain the datasets
    # Download them if they are not present
    train_dataset = torchvision.datasets.CIFAR10(root=datapath,
train=True,
    transform=train_transforms, download=True)
    test_dataset = torchvision.datasets.CIFAR10(root=datapath,
train=False,
    transform=test_transforms, download=True)

    # Create Data Loaders
    train_loader = torch.utils.data.DataLoader(train_dataset, batch_
size=128,
    shuffle=True, num_workers=2)
    test_loader = torch.utils.data.DataLoader(test_dataset, batch_
size=128,
    shuffle=False, num_workers=2)

    return train_loader, test_loader
```

PyTorch provides a standard functionality known as data loaders that facilitates easy data manipulation for ML. Data loaders can generate the data needed by applying specific transformations, and the generated data can be consumed by ML models. Note that when defining the data loader, we have specified a `batch_size` parameter. This defines the number of data instances that will be read at a time. In our case, it is set to `128`, which means that the forward pass, loss calculation, backpropagation, and gradient descent will happen one by one for batches where each batch is of 128 images.

Next, we will train a vanilla NN model for image classification on the CIFAR dataset. We first perform some boilerplate setup that includes the following:

1. Setting the number of epochs to be used for training.
2. Loading the train and test data loaders.
3. Initializing the model with the basic CNN model we defined earlier.
4. Defining the loss function to be cross-entropy and the optimizer to be Adam.
5. Moving the model to CUDA if it is available.

Then, we begin the training loop. In each iteration of the training loop, we do the following:

1. Load a batch of training data.
2. Move it to the GPU (CUDA) if needed and available.
3. Zero out the optimizer gradients.
4. Calculate the predicted output by running inference on the model.
5. Calculate the loss based on prediction and ground truth.
6. Backpropagate the loss.

Here is a code snippet that executes these steps one by one:

```
NUM_EPOCHS = 10

train_data, test_data = load_cifar10_datasets(datapath = "./data")
model = BasicImageNetCNN(in_channels = 3)
loss_function = torch.nn.CrossEntropyLoss(reduction="mean")
optimizer = torch.optim.Adam(model.parameters(), lr=1e-3)

if torch.cuda.is_available():
  device = "cuda"
  model = model.cuda()
else:
  device = "cpu"

model.train()

for epoch in range(NUM_EPOCHS):
  train_loss = 0.0
```

```
for x, y in train_data:

    # Move image and labels to device if applicable
    x = x.to(device)
    y = y.to(device)

    # Zero out the gradients from previous epoch if any
    optimizer.zero_grad()

    # Calculate predicted value and loss
    y_pred = model(x)
    loss = loss_function(y_pred, y)

    # Backpropagation
    loss.backward()
    optimizer.step()

    # Keep track of the loss
    train_loss = train_loss + loss.item()

    # Print some information for logging
    print("EPOCH: {} ---------- Loss: {}".format(epoch, train_loss))
```

Finally, we evaluate our model. While evaluating, we evaluate the model on two sets of data—the original test data and the adversarial test data that we created using FGSM.

We first set the model to evaluation mode, which means that gradients are not computed and stored. This makes the operations more efficient, as we do not need the overhead of gradients during inference on the model. For every batch of the training data, we calculate the adversarial images using FGSM. Here, we have set the value of epsilon to 0.005. Then, we run inference on the model using both the clean images (the original test set) and adversarial images (generated through FGSM). For every batch, we will calculate the number of examples for which the predicted and ground-truth labels match, which will give us the accuracy of the model. Comparing the accuracy of the clean and adversarial set shows us how effective our adversarial attack is:

```
model.eval()
clean_correct = 0
fgsm_correct = 0
total = 0
for x, y in test_data:

    # Move image and labels to device if applicable
    x = x.to(device)
```

```
    y = y.to(device)

    # Calculate the adversarial images
    x_fgsm = Generate_FGSM_Image(model, x, epsilon = 0.005)

    # Run inference for predicted values on clean and adversarial
examples
    _, y_pred_clean = torch.max(model(x), 1)
    _, y_pred_fgsm = torch.max(model(x_fgsm), 1)

    # Calculate accuracy of clean and adversarial predictions
    clean_correct = clean_correct + y_pred_clean.eq(y).sum().item()
    fgsm_correct = fgsm_correct + y_pred_fgsm.eq(y).sum().item()
    total = total + y.size(0)

clean_accuracy = clean_correct / total
fgsm_accuracy = fgsm_correct / total
```

This concludes our discussion of the FGSM attack. You can compare the accuracy before and after the adversarial perturbation (`clean_accuracy` and `fgsm_accuracy`, respectively). The drop in accuracy indicates the effectiveness of the adversarial attack.

PGD

In the previous section, we discussed the FGSM attack method and how it can be used to generate adversarial images by adding small perturbations to the input image based on the sign of the gradient. The **PGD** method extends FGSM by applying it iteratively.

PGD working

Specifically, the PGD attack will, for an input image, calculate a perturbation based on the FGSM attack. Adding this to the image will give us the perturbed image. Whereas the FGSM attack stops here, the PGD attack goes a step further. Once an adversarial image has been generated, we clip the image. Clipping refers to adjusting the image so that it remains in the neighborhood of the original image. Clipping is done on a per-pixel basis. After the image has been clipped, we repeat the process multiple times iteratively to obtain the final adversarial image.

Formally speaking, given an image X, a series of adversarial images can be calculated as follows:

$$X_{N+1}' = Clip_{X,\epsilon} \left(X_N' + \alpha \cdot sign \left(\nabla_x \mathscr{L}(\theta, x, y) \right) \right)$$

The notation here is slightly different from that for the FGSM attack. Here, α serves the same role as ϵ did in FGSM; it controls the amount of perturbation. Typically, it is set to 1, which means that each pixel is modified by at most one unit in each step. The process is repeated iteratively for a predetermined number of steps.

The function that implements this is quite straightforward. It simply uses FGSM iteratively and clips the generated images. The FGSM function must be modified to take in the predicted ground-truth label by the model, as it will change in every step and should not be recalculated by FGSM. So, we pass it the ground truth as a parameter and use that instead of recalculating it as a model prediction. In the FGSM function, we simply use the value that is passed in instead of running inference on the model.

The modified FGSM function is shown as follows:

```
def Generate_FGSM_Image_V2(model,x,y, // New Parameter.epsilon):

    # Check if epsilon is 0
    # If so, that means no perturbation is added
    # We can avoid gradient calculations
    if epsilon == 0:
      return x

    # Convert x to a float and having gradients enabled
    x = x.clone().detach()
    x = x.to(torch.float)
    x - x.requires_grad_(True)

    # Compute Loss
    loss_function = nn.CrossEntropyLoss()
    loss = loss_function(model(x), y)

    # Backpropagate Loss
    loss.backward()

    # Calculate perturbation using the FGSM equation
    perturbation = epsilon * torch.sign(x.grad)

    # Calculate the adversarial image
    x_adversarial = x + perturbation

    return x_adversarial
```

For every image, the PGD method attack function completes the following steps:

1. Calculates the predicted label by running inference on the model.

2. Sets the original image as the initial adversarial image X_0.

3. Calculates the adversarial image using the FGSM attack method described in the previous section. In doing so, it passes the predicted value as a parameter so that FGSM does not recompute it in every cycle.

4. Computes the difference between the image and the adversarially generated image. This is the perturbation to be added.

5. Clips this perturbation so that the adversarial image is within the neighborhood of the original image.

6. Adds the clipped perturbation to the image to obtain the adversarial image.

7. Repeats *steps 2-6* for the desired number of iterations to obtain the final adversarial image.

As you can see, the overarching idea remains the same as with FGSM, but only the process of generating the adversarial images changes.

PGD implementation

A code snippet for the PGD method is shown as follows:

```
def Generate_PGDM_Image(model,x,epsilon,num_iterations):

    # Obtain actual clean predictions from model
    _, y = torch.max(model(x), 1)

    # Calculate the initial adversarial value
    eta = torch.zeros_like(x)
    eta = torch.clamp(eta, -1*eps, 1*eps)
    x_adv = x + eta

    # For every iteration, do FGSM and clipping
    for _ in range(num_iterations):

        # Note that the FGSM function signature has changed
        # We are passing it the predicted value y as a parameter
        # Thus this will not be recomputed
        x_adv = Generate_FGSM_Image_V2(model,x_adv,y,epsilon = 0.01)

        eta = x_adv - x
        eta = torch.clamp(eta, -1*eps, 1*eps)
        x_adv= x + eta

    # Return the final image
    return x_adv
```

This function can be used to generate adversarial images given an image using the PGD method. We will not repeat the experiment of model setup, training, and evaluation. Simply using the Generate_PGDM_Image() function instead of the Generate_FGSM_Image() function should allow you to run our analysis using this attack. How does the performance of this attack compare to the FGSM attack?

This concludes our discussion of attacking image models. In the next section, we will discuss attacking text models.

Attacking text models

Please note that this section contains examples of hate speech and racist content online.

Just as with images, text models are also susceptible to adversarial attacks. Attackers can modify the text so as to trigger a misclassification by ML models. Doing so can allow an adversary to escape detection.

A good example of this can be seen on social media platforms. Most platforms have rules against abusive language and hate speech. Automated systems such as keyword-based filters and ML models are used to detect such content, flag it, and remove it. If something outrageous is posted, the platform will block it at the source (that is, not allow it to be posted at all) or remove it in the span of a few minutes.

A malicious adversary can purposely manipulate the content in order to fool a model into thinking that the words are out of vocabulary or are not certain abusive words. For example, according to a study (*Poster | Proceedings of the 2019 ACM SIGSAC Conference on Computer and Communications Security* (`https://dl.acm.org/doi/abs/10.1145/3319535.3363271`)), the attacker can manipulate their post as shown:

Figure 9.3 – A hateful tweet and the adversarially manipulated version

The original tweet that says "*go back to where you came from -- these fucking immigrants are destroying America!!!*" is clearly hate speech and racism against immigrants. Originally, this was classified to be 95% toxic, that is, a toxicity classifier assigned it the label *toxic* with 95% probability. Obviously, this classification is correct.

In the obfuscated tweet, the attacker modified three words by eliminating one letter from three words. Note that we still very much recognize those words for what they are. The intent is clear, and this is still very much hate speech. An automated system, however, will not know this. Models and rules work by looking at words. To them, these new words are out of their vocabulary. They have not seen the words *imigrants*, *fuckng*, or *destroyin* in prior examples of hate speech during training. Therefore, the model misclassifies it and assigns it a label of not being toxic content with a probability of 63%. The attacker thus succeeded in fooling a classifier to pass off their toxic content as benign.

The principle of adversarial attacks in text is the same as that of those in images: manipulating the input so as to confuse the model and not allowing it to recognize certain important features. However, two key differences set adversarial manipulation in text apart from adversarial manipulation in images.

First, the adversarially generated input should be reasonably similar to the original input. For example, we saw in the preceding section that the two images of the panda were nearly identical. This will not be the case with text—the changes made through manipulation will be visible and discernible to the naked eye. Looking at the screenshot of tweets that we just discussed, we can clearly see that the words are different. Manipulation in text is obvious whereas that in images is not. As a result, there is a limit to how much we can manipulate. We cannot change every word—too many changes will clearly indicate the manipulation and lead to discovery, thus defeating the goal.

Second, images are robust to change as compared to text. Changing multiple pixels in an image will still leave the larger image mainly unchanged (that is, a panda will still be recognizable, maybe with some distortion). On the other hand, text depends on words for the meaning it provides. Changing a few words will change the meaning entirely, or render the text senseless. This would be unacceptable—the goal of an adversarial attack is to still have the original meaning.

Manipulating text

In this section, we will explore techniques for manipulating text so that it may fool an ML classifier. We will show a few sample techniques and provide general guidelines at the end for further exploration.

Recall the example of hate speech online that we discussed earlier: attackers can manipulate text so as to escape detection and post toxic content online. In this section, we will attempt to build such techniques and examine whether we can beat ML models.

Data

For this experiment, we will use the *Toxic Tweets* dataset (*Toxic Tweets Dataset | Kaggle*) (https://www.kaggle.com/datasets/ashwiniyer176/toxic-tweets-dataset). This data is made available freely as part of a Kaggle challenge online. You will have to download the data, and then unzip it to extract the CSV file. The data can then be read as follows:

```
import pandas as pd
import numpy as np
```

```
df = pd.read_csv("FinalBalancedDataset.csv", skiprows = 1, names=
["TweetId","Toxicity","Tweet"])
df.head()
```

This should show you what the data looks like, as follows:

	TweetId	Toxicity	Tweet
0	0	0	@user when a father is dysfunctional and is s...
1	1	0	@user @user thanks for #lyft credit i can't us...
2	2	0	bihday your majesty
3	3	0	#model i love u take with u all the time in ...
4	4	0	factsguide: society now #motivation

Figure 9.4 – Hate speech dataset

You can also look at the distribution of the labels as follows:

```
df.groupby("Toxicity").count()["TweetId"]
```

This will show you the following output:

```
Toxicity
0    32592
1    24153
Name: TweetId, dtype: int64
```

Figure 9.5 – Distribution of tweets by toxicity label

The dataset contains approximately 24,000 tweets that are toxic and 32,500 that are not. In the next section, we will extract features from this data.

Extracting features

In our earlier chapters, we discussed that there needs to be a method to extract features from text, and such features must be numeric in value. One such method, which we have already used, is the **Term Frequency-Inverse Document Frequency (TF-IDF)** approach. Let us do a brief recap of TF-IDF.

TF-IDF is a commonly used technique in **natural language processing (NLP)** to convert text into numeric features. Every word in the text is assigned a score that indicates how important the word is in that text. This is done by multiplying two metrics, as follows:

- **TF**: How frequently does the word appear in the text sample? This can be normalized by the length of the text in words, as texts that differ in length by a large number can cause skews. The TF metric measures how common a word is in this particular text.

- **IDF**: How frequently does the word appear in the rest of the corpus? First, the number of text samples containing this word is obtained. The total number of samples is divided by this number. Simply put, IDF is the inverse of the fraction of text samples containing the word. The IDF metric measures how common the word is in the rest of the corpus.

More details on TF-IDF can be found in *Chapter 7, Attributing Authorship and How to Evade It*. For now, here is a code snippet to extract the TF-IDF features for a list of sentences:

```
from sklearn.feature_extraction.text import TfidfVectorizer
def Extract_TF_IDF(train_data, test_data):

    tf_idf = TfidfVectorizer()
    X_train_TFIDF = tf_idf.fit_transform(train_data)
    X_test_TFIDF = tf_idf.transform(test_data)

    return X_train_TFIDF, X_test_TFIDF
```

Adversarial attack strategies

We will attempt to evade our ML models using two adversarial strategies: appending a letter at the end of some words, and repeating certain vowels from some words. In each case, our end goal is to fool the classifier into thinking that there are words that it has not seen and, therefore, it does not recognize those words. Let us discuss these strategies one by one.

Doubling the last letter

In this strategy, we simply misspell the word by appending an additional letter at the end. We double the last letter so that the word appears to be unchanged, and is still recognizable. For example, *America* will become *Americaa*, and *immigrant* will become *immigrantt*. To a machine, these words are totally different from one another.

Here is the code to implement this:

```
def double_last_letter(sentences, max_perturbations = 3):

    # Output array
    modified_sentences = []
```

```
    for sentence in sentences:

        # Split into words
        words = sentence.split(' ')

        # Randomly choose words to manipulate
        rand_indices = np.random.randint(0, len(words), max_
perturbations)

        for idx in rand_indices:

            # Check if the word is blank, if yes, skip
            if len(words[idx]) == 0:
                continue

            # Double the last letter in the chosen word
            words[idx]+=words[idx][-1]

        # Join back to make sentence
        modified_sentences.append(' '.join(word for word in words))

    return modified_sentences
```

Doubling vowels

In this attack, we will look for words with vowels, and on finding the first vowel, we will repeat it. For example, *Facebook* will become *Faacebook*, and *Coronavirus* will become *Cooronavirus*. It is fairly intuitive that repeated vowels often go unnoticed when reading text; this means that the text will appear to be unchanged to a quick reader. The following code snippet implements this attack:

```
def double_vowel(sentences, max_perturbations = 3):

    total_perturbations = 0
    # Output array
    modified_sentences = []

    for sentence in sentences:

        # Split into words
        words = sentence.split(' ')

        for i in range(len(words)):
```

```
        # Check if maximum perturbations done
        # If so, break the loop and don't do any more!
        if total_perturbations>max_perturbations:
            break

        for vowel in ['a','e','i','o','u']:
            if vowel in words[i]:
                words[i] = words[i].replace(vowel,vowel+vowel,1)
                total_perturbations+=1

                # Here replace only for one vowel
                # So once replacement is done, break out
                # This will break only this loop
                break

    modified_sentences.append(' '.join(word for word in words))

    return modified_sentences
```

Executing the attacks

Now that we have defined the two attacks we will implement, it is time to actually execute them. To achieve this, we will do the following:

1. Split the data into training and test sets.

2. Build a TF-IDF model on the training data and use it to extract features from the training set.

3. Train a model based on the features extracted in *step 2*.

4. Use the same TF-IDF model to extract features on the test set and run inference on the trained model.

5. Calculate metrics for classification—accuracy, precision, recall, and F-1 score. These are the baseline scores.

6. Now, apply the attack functions and derive the adversarial test set.

7. Use the same TF-IDF model to extract features from the adversarial test set and run inference on the trained model.

8. Calculate metrics for classification—accuracy, precision, recall, and F-1 score. These are the scores upon adversarial attack.

Comparing the scores obtained in *steps 5* and *8* will tell us what the effectiveness of our attack was.

First, we split the data and extract features for our baseline model:

```
X = df["Tweet"].tolist()
y = df["Toxicity"].tolist()

from sklearn.model_selection import train_test_split
X_train, X_test, Y_train, Y_test = train_test_split(X,y,
test_size = 0.3,stratify = y)

X_train_features, X_test_features = Extract_TF_IDF(X_train, X_test)
```

Let us also set up an evaluation function that takes in the actual and predicted values and prints out our metrics, such as accuracy, precision, recall, and F-1 score:

```
from sklearn.metrics import confusion_matrix

def evaluate_model(actual, predicted):
  confusion = confusion_matrix(actual, predicted)
  tn, fp, fn, tp = confusion.ravel()

  total = tp + fp + tn + fn

  accuracy = 1.0 * (tp + tn) / total
  if tp + fp != 0:
    precision = tp / (tp + fp)
  else:
    precision = 0

  if tp + fn != 0:
    recall = tp / (tp + fn)
  else:
    recall = 0

  if precision == 0 or recall == 0:
    f1 = 0
  else:
    f1 = 2 * precision * recall / (precision + recall)

  evaluation = { 'accuracy': accuracy,
                 'precision': precision,
                 'recall': recall,
                 'f1': f1}

  return evaluation
```

Now, we build and evaluate our baseline model. This model has no adversarial perturbation:

```
from sklearn.ensemble import RandomForestClassifier
model = RandomForestClassifier(n_estimators = 100)
model.fit(X_train_features, Y_train)
Y_predicted = model.predict(X_test_features)
evaluation = evaluate_model(Y_test, Y_predicted)

print("Accuracy: {}".format(str(evaluation['accuracy'])))
print("Precision: {}".format(str(evaluation['precision'])))
print("Recall: {}".format(str(evaluation['recall'])))
print("F-1: {}".format(str(evaluation['f1'])))
```

This should result in an output as shown:

```
Accuracy: 0.9302161654135338
Precision: 0.9291584018135449
Recall: 0.905051062655258
F-1: 0.9169463087248322
```

Figure 9.6 – Metrics for classification of normal data

Next, we actually execute the attack. We obtain adversarial samples but train the model on the clean data (as during training we do not have access to the attacker's adversarial set). Here, we use the `double_last_letter()` adversarial function to compute our adversarial set. We then evaluate the model on the adversarial samples:

```
# Obtain adversarial samples
X_test_adversarial = double_last_letter(X_test, max_perturbations=5)

# Extract features
X_train_features, X_test_features = Extract_TF_IDF(X_train, X_test_
adversarial)

# Train model
model = RandomForestClassifier(n_estimators = 100)
model.fit(X_train_features, Y_train)

# Predict on adversarial samples
Y_predicted = model.predict(X_test_features)

# Evaluate
evaluation = evaluate_model(Y_test, Y_predicted)
print("Accuracy: {}".format(str(evaluation['accuracy'])))
print("Precision: {}".format(str(evaluation['precision'])))
print("Recall: {}".format(str(evaluation['recall'])))
print("F-1: {}".format(str(evaluation['f1'])))
```

This should show you another set of metrics:

```
Accuracy: 0.8774083646616542
Precision: 0.9141114143522234
Recall: 0.785812862268838
F-1: 0.8451205936920222
```

Figure 9.7 – Metrics for classification of adversarial data

Carefully note the differences between the scores obtained with the clean and adversarial data; we will compare them side by side for clarity, as follows:

Metric	Clean	Adversarial
Accuracy	0.93	0.88
Precision	0.93	0.91
Recall	0.90	0.79
F-1 score	0.92	0.85

Table 9.1 – Comparing accuracy of normal versus adversarial data

You can clearly see that our adversarial attack has been successful—while accuracy dropped only by 5%, the recall dropped by 11%, causing a 7% drop in the F-1 score. If you consider this at the scale of a social media network such as Twitter, it translates to significant performance degradation.

You can similarly evaluate the effect caused by the double-vowel attack. Simply generate adversarial examples using the double-vowel function instead of the double-last letter function. We will leave this as an exercise to the reader.

Further attacks

The previous section covered only two basic attacks for attacking text-based models. As a data scientist, you must think of all possible attack surfaces and come up with potential new attacks. You are encouraged to develop and implement new attacks and examine how they affect model performance. Some potential attacks are presented here:

- Combining two words from the sentence (for example, *Live and Let Live* will become *Liveand Let Live*)

- Splitting a long word into two words (for example, *Immigrants will take away our jobs* will become *Immi grants will take away our jobs*)

- Adding hyphens or commas to long words (for example, *Immigrants will take away our jobs* will become *Im-migrants will take away our jobs*)

Additionally, readers should also experiment with different feature extraction methods to examine whether any of them is more robust to adversarial attacks than the others, or our TF-IDF method. A few examples of such methods are set out here:

- Word embeddings
- Contextual embeddings (**Bidirectional Encoder Representations from Transformers**, or **BERT**)
- Character-level features

This completes our discussion on how text models can be fooled. In the next section, we will briefly discuss how models can be made robust against adversarial attacks.

Developing robustness against adversarial attacks

Adversarial attacks can be a serious threat to the security and reliability of ML systems. Several techniques can be used to improve the robustness of ML models against adversarial attacks. Some of these are described next.

Adversarial training

Adversarial training is a technique where the model is trained on adversarial examples in addition to the original training data. Adversarial examples are generated by perturbing the original input data in such a way that the perturbed input is misclassified by the model. By training the model on both the original and adversarial examples, the model learns to be more robust to adversarial attacks. The idea behind adversarial training is to simulate the types of attacks that the model is likely to face in the real world and make the model more resistant to them.

Defensive distillation

Defensive distillation is a technique that involves training a model on soft targets rather than hard targets. Soft targets are probability distributions over the classes, while hard targets are one-hot vectors indicating the correct class. By training on soft targets, the decision boundaries of the model become smoother and more difficult to attack. This is because the soft targets contain more information about the distribution of the classes, which makes it more difficult to create an adversarial example that will fool the model.

Gradient regularization

Gradient regularization is a technique that involves adding a penalty term to the loss function of the model that encourages the gradients of the model to be small. This helps to prevent an attacker from creating an adversarial example by perturbing the input in the direction of the gradient. The penalty term can be added to the loss function in various ways, such as through L1 or L2 regularization or by using adversarial training. Gradient regularization can be combined with other techniques to improve the robustness of the model.

Input preprocessing

Input preprocessing involves modifying the input data before it is fed into the model. This can include techniques such as data normalization, which helps to reduce the sensitivity of the model to small changes in the input. Other techniques include randomization of the input, which can help to disrupt the pattern of the adversarial attack, or filtering out anomalous input that may be indicative of an adversarial attack. Input preprocessing can be tailored to the specific model and the type of input data that it receives.

Ensemble methods

Ensemble methods involve combining multiple models to make a prediction. This can improve the robustness of the model by making it more difficult for an attacker to craft an adversarial example that will fool all the models in the ensemble. Ensemble methods can be used in conjunction with other techniques to further improve the robustness of the model.

Certified defenses

Certified defenses involve creating a provable guarantee that a model will be robust to a certain level of adversarial attack. This can be done using techniques such as interval-bound propagation or randomized smoothing. Interval-bound propagation involves computing a range of values that the model's output can take given a certain range of inputs. This range can be used to create a provable bound on the robustness of the model. Randomized smoothing involves adding random noise to the input data to make the model more robust to adversarial attacks. Certified defenses are a relatively new area of research, but hold promise for creating more robust ML models.

It's worth noting that while these techniques can improve the robustness of ML models against adversarial attacks, they are not foolproof, and there is still ongoing research in this area. It's important to use multiple techniques in combination to improve the robustness of the model.

With that, we have come to the end of this chapter.

Summary

In recent times, human reliance on ML has grown exponentially. ML models are involved in several security-critical applications such as fraud, abuse, and other kinds of cybercrime. However, many models are susceptible to adversarial attacks, where attackers manipulate the input so as to fool the model. This chapter covered the basics of AML and the goals and strategies that attackers employ. We then discussed two popular adversarial attack methods, FGSM and PGD, along with their implementation in Python. Next, we learned about methods for manipulating text and their implementation.

Because of the importance and prevalence of ML in our lives, it is necessary for security data scientists to understand adversarial attacks and learn to defend against them. This chapter provides a solid foundation for AML and the kinds of attacks involved.

So far, we have discussed multiple aspects of ML for security problems. In the next chapter, we will pivot to a closely related topic—privacy.

10

Protecting User Privacy with Differential Privacy

With the growing prevalence of machine learning, some concerns have been raised about how it could potentially be a risk to user privacy. Prior research has shown that even carefully anonymized datasets can be analyzed by attackers and de-anonymized using pattern analysis or background knowledge. The core idea that privacy is based upon is a user's right to control the collection, storage, and use of their data. Additionally, privacy regulations mandate that no sensitive information about a user should be leaked, and they also restrict what user information can be used for machine learning tasks such as ad targeting or fraud detection. This has led to concerns about user data being used for machine learning, and privacy is a crucial topic every data scientist needs to know about.

This chapter covers differential privacy, a technique used to perform data analysis while maintaining user privacy at the same time. Differential privacy aims to add noise to data and query results, such that the query accuracy is retained but no user data is leaked. This can help with simple analytical tasks as well as machine learning. We will start by understanding the fundamentals of privacy and what it means for users, and for you as engineers, scientists, and developers. We will then also work our way through privacy by design, and the legal implications of violating privacy regulations. Finally, we will implement differential privacy in machine learning and deep learning models.

In this chapter, we will cover the following main topics:

- The basics of privacy
- Differential privacy
- Differentially private machine learning
- Differentially private deep learning

By the end of this chapter, you will have a better understanding of why privacy is important and how it can be incorporated into machine learning systems.

Technical requirements

You can find the code files for this chapter on GitHub at `https://github.com/PacktPublishing/10-Machine-Learning-Blueprints-You-Should-Know-for-Cybersecurity/tree/main/Chapter%2010`.

The basics of privacy

Privacy is the ability of an individual or a group of individuals to control their personal information and to be able to decide when, how, and to whom that information is shared. It involves the right to be free from unwanted or unwarranted intrusion into their personal life and the right to maintain the confidentiality of personal data.

Privacy is an important aspect of individual autonomy, and it is essential for maintaining personal freedom, dignity, and trust in personal relationships. It can be protected by various means, such as legal safeguards, technological measures, and social norms.

With the increasing use of technology in our daily lives, privacy has become an increasingly important concern, particularly in relation to the collection, use, and sharing of personal data by organizations and governments. As a result, there has been growing interest in developing effective policies and regulations to protect individual privacy. In this section, we will cover the fundamental concepts of privacy, associated legal measures, and the implications to machine learning.

Core elements of data privacy

The fundamental principle behind data privacy is that users should be able to answer and have control over the following questions:

- What data about me is being collected?
- What will that data be used for?
- Who will have access to the data?
- How will the data be protected?

In this section, we will explore these concepts in detail.

Data collection

Data collection refers to the process of gathering personal information or data from individuals. This data can include any information that can identify an individual, such as name, address, phone number, email, date of birth, social security number, and so on. Organizations that collect personal data must ensure that the data is collected only for specific, legitimate purposes and that individuals are made aware of what data is being collected and why. In the case of fraud detection or other security applications, data collection may seem overly intrusive (such as private messages being collected for detecting abuse, or computer processes for detecting malware). Additionally, data collection must comply with applicable laws and regulations, and organizations must obtain explicit consent from individuals before collecting their data.

Data use

Data use refers to how the collected data is used by organizations or individuals. Organizations must ensure that they use personal data only for the specific, legitimate purposes for which it was collected and that they do not use the data for any other purposes without the individual's explicit consent. Additionally, organizations must ensure that they do not use personal data in a way that discriminates against individuals, such as denying them services or opportunities based on their personal characteristics. Data use also includes using the data for machine learning models as training – some users may not want their data to be used for training or analysis.

Data access

Data access refers to the control that individuals have over their personal data. Individuals have the right to know what data is being collected about them, who is collecting it, and why it is being collected. They also have the right to access their own personal data and correct any inaccuracies. Additionally, individuals have the right to know who their data is being shared with and for what purposes. This also includes data sharing with other organizations, applications, and services (for example, a shopping website selling your search history with a marketing company). Personal data should only be shared with the individual's explicit consent and should only be shared for specific, legitimate purposes. Organizations must ensure that they have appropriate security measures in place to protect personal data from unauthorized access, disclosure, alteration, or destruction.

Data protection

Data protection refers to the measures taken to protect personal data from unauthorized access, disclosure, alteration, or destruction. This includes technical, physical, and administrative measures to ensure the security and confidentiality of personal data. Organizations must ensure that they have appropriate security measures in place to protect personal data, such as encryption, access controls, and firewalls. Additionally, organizations must ensure that they have policies and procedures in place to detect and respond to security incidents or breaches.

Privacy and the GDPR

While privacy has to do with user consent on data, it is not a purely ethical or moral concern – there are legal requirements and regulations that organizations must comply with. The **GDPR** stands for the **General Data Protection Regulation**. It is a comprehensive data protection law that came into effect on May 25, 2018, in the **European Union** (**EU**).

The GDPR regulates the processing of personal data of individuals within the EU, as well as the export of personal data outside the EU. It gives individuals more control over their personal data and requires organizations to be transparent about how they collect, use, and store personal data.

The GDPR sets out several key principles, including the following:

- Lawfulness, fairness, and transparency
- Purpose limitation
- Data minimization
- Accuracy
- Storage limitation
- Integrity and confidentiality (security)
- Accountability

Under the GDPR, individuals have the right to access their personal data, correct inaccurate data, have their data erased in certain circumstances, and object to the processing of their data. Organizations that fail to comply with the GDPR can face significant fines and other sanctions.

For example, in January 2019, the CNIL (the French data protection authority) fined Google €50 million for GDPR violations related to the company's ad personalization practices. The CNIL found that Google had violated the GDPR in two key ways:

- **Lack of transparency**: The CNIL found that Google had not provided users with clear and easily accessible information about how their personal data was being used for ad personalization. The information was spread across several different documents, making it difficult for users to understand.
- **Lack of valid consent**: The CNIL found that Google had not obtained valid consent from users for ad personalization. The consent was not sufficiently informed, as users were not clearly told what specific data was being collected and how it was being used.

The CNIL's investigation was initiated following two complaints filed by privacy advocacy groups, **None Of Your Business (NOYB)** and La Quadrature du Net, on May 25, 2018, the same day that the GDPR came into effect.

In addition to the fine, the CNIL ordered Google to make changes to its ad personalization practices, including making it easier for users to access and understand information about how their data is being used and obtaining valid consent.

The Google fine was significant, as it was the largest GDPR fine at the time and demonstrated that regulators were willing to take enforcement action against large tech companies for GDPR violations. The fine also underscored the importance of transparency and valid consent in data processing under the GDPR.

The GDPR has had a significant impact on how organizations handle personal data, not only in the EU but also worldwide, as many companies have had to update their policies and practices to comply with the regulation.

Privacy by design

Privacy by design is an approach to privacy protection that aims to embed privacy and data protection into the design and architecture of systems, products, and services from the outset. The concept was first introduced by the Information and Privacy Commissioner of Ontario, Canada, in the 1990s, and has since been adopted as a best practice by privacy regulators and organizations worldwide.

The privacy-by-design approach involves proactively identifying and addressing privacy risks, rather than trying to retrofit privacy protections after a system or product has been developed. It requires organizations to consider privacy implications at every stage of the design process, from the initial planning and conceptualization phase through to implementation and ongoing operation.

As data scientists and machine learning engineers, if you are designing any system at scale, understanding privacy concerns is important. You should follow the principles of privacy by design while developing any system. There are five key principles that define privacy by design: proactivity, privacy as default, privacy embedded into the design, full functionality, and end-to-end security.

Proactive not reactive

The principle of being proactive not reactive means that organizations should anticipate potential privacy risks and take steps to mitigate them before they become a problem. This involves conducting **privacy impact assessments** (**PIAs**) to identify and address potential privacy issues at the outset of a project. By taking a proactive approach to privacy, organizations can reduce the likelihood of privacy breaches, protect individual rights, and build trust with their customers.

Privacy as the default setting

The principle of privacy as the default setting means that individuals should not have to take any action to protect their privacy. This means that privacy protection should be built into systems, products, and services by default and that individuals should not be required to opt out of sharing their data. By making privacy the default setting, individuals are empowered to make informed decisions about their personal information, without having to navigate complex privacy settings or policies.

Privacy embedded into the design

The principle of embedding privacy into the design means that privacy should be a core consideration in the development of systems, products, and services from the outset. This involves incorporating privacy features and controls into the design of these products and services, such as anonymization, encryption, and data minimization. By building privacy into the design of products and services, organizations can help ensure that privacy is protected by default, rather than as an afterthought.

Full functionality

The principle of full functionality means that privacy protections should not come at the expense of functionality or usability. This means that privacy protections should be integrated into systems and products without compromising their performance or functionality. By adopting a positive-sum approach to privacy, organizations can build trust with their customers and demonstrate that they take privacy seriously.

End-to-end security

The principle of end-to-end security means that comprehensive security measures should be implemented throughout the entire life cycle of a product or service, from development to disposal. This involves implementing a range of security measures, such as access controls, encryption, and monitoring, to protect against unauthorized access, use, and disclosure of personal information. By taking a comprehensive approach to security, organizations can help ensure that personal information is protected at every stage of the life cycle and build trust with their customers.

Privacy and machine learning

Why did we spend all this time discussing the concepts behind privacy, the elements of data privacy, and the GDPR? In security areas (such as fraud, abuse, and misinformation), significant types and amounts of user data are collected. Some of it might be deemed obtrusive, such as the following:

- Browsers collecting users' mouse movements and click patterns to detect click fraud and bots
- Security software collecting information on system processes to detect the presence of malware
- Social media companies extracting information from private messages and images to detect child pornography

For data scientists in the security domain, the ultimate goal is to provide maximum user security by building a system with the highest precision and recall. However, at the same time, it is important to understand the limitations you will face in data collection and use while designing your systems. For example, if you are building a system for fraud detection, you may not be able to use cookie data in France. Additionally, the GDPR will apply if the data you are collecting is from European users.

Depending on the jurisdiction, you may not be able to collect certain data, or even if it is collected, you may not be able to use it for machine learning models. These factors must be taken into consideration as you design your systems and algorithms.

Furthermore, we know that machine learning is based on identifying trends and patterns from data. Privacy considerations and regulations will severely limit your ability to collect data, extract features, and train models.

Now that you have been introduced to the fundamentals of privacy, we will look at differential privacy, which is considered to be state-of-the-art in privacy and is used by many tech giants.

Differential privacy

In this section, we will cover the basics of differential privacy, including the mathematical definition and a real-world example.

What is differential privacy?

Differential privacy (DP) is a framework for preserving the privacy of individuals in a dataset when it is used for statistical analysis or machine learning. The goal of DP is to ensure that the output of a computation on a dataset does not reveal sensitive information about any individual in the dataset. This is accomplished by adding controlled noise to the computation in order to mask the contribution of any individual data point.

DP provides a mathematically rigorous definition of privacy protection by quantifying the amount of information that an attacker can learn about an individual by observing the output of a computation. Specifically, DP requires that the probability of observing a particular output from a computation is roughly the same whether a particular individual is included in the dataset or not.

Formally speaking, let D and D' be two datasets that differ by, at most, one element, and let f be a function that takes a dataset as input and produces an output in some range, R. Then, the f function satisfies ε-differential privacy for any two datasets D and D':

$$\Pr[f(D) \in S] \leq e^{\varepsilon} \cdot \Pr[f(D') \in S]$$

where S is any subset of R and δ is a small positive number that accounts for the probability of events that have low probability. In other words, the probability of the output of the f function on dataset D falling within a set S should be very similar to the probability of the output of the f function on dataset D' falling within the same set S, up to a multiplicative factor of $\exp(\varepsilon)$. The smaller the value of ε, the stronger the privacy protection, but also the less accurate the results. The δ parameter is typically set to a very small value, such as 10^{-9}, to ensure that the overall privacy guarantee is strong.

The key idea behind DP is to add random noise to the computation in a way that preserves the statistical properties of the data while obscuring the contribution of any individual. The amount of noise added is controlled by a parameter called the **privacy budget**, which determines the maximum amount of privacy loss that can occur during the computation.

There are several mechanisms for achieving differential privacy, including the following:

- **Laplace mechanism**: The Laplace mechanism adds random noise to the output of a computation based on the sensitivity of the computation. The amount of noise added is proportional to the sensitivity of the computation and inversely proportional to the privacy budget.

- **Exponential mechanism**: The Exponential mechanism is used to select an output from a set of possible outputs in a way that minimizes the amount of information revealed about any individual. This mechanism selects the output with the highest utility score, where the utility score is a measure of how well the output satisfies the desired properties of the computation.

- **Randomized response:** Randomized response is a mechanism used to obtain accurate estimates of binary data while preserving privacy. The mechanism involves flipping the value of the data point with a certain probability, which is determined by the privacy budget.

DP has become increasingly important in recent years due to the widespread use of data in machine learning and statistical analysis. DP can be used to train machine learning models on sensitive data while ensuring that the privacy of individuals in the dataset is preserved. It is also used in other applications, such as census data and medical research.

Differential privacy – a real-world example

The concept of differential privacy can be clarified in detail using a practical example. Suppose a credit card company has a dataset containing information about the transaction amounts and times of its customers, and they want to identify potential cases of fraud. However, the credit card company is concerned about preserving the privacy of its customers and wants to ensure that the analysis cannot be used to identify the transaction amounts or times of any individual customer.

To accomplish this, the credit card company can use differential privacy to add noise to the analysis. Specifically, they can add random noise to the computed statistics in a way that preserves the overall statistical properties of the data but makes it difficult to determine the transaction amounts or times of any individual customer.

For example, the credit card company could use the Laplace mechanism to add noise to the computed statistics. Let's say the credit card company wants to compute the total transaction amount for a specific time period, and the sensitivity of the computation is *1*, meaning that changing the amount of one transaction can change the computed total by, at most, 1 dollar. The credit card company wants to achieve a privacy budget of *epsilon = 1*, meaning that the probability of observing a particular output from the computation should be roughly the same whether a particular customer is included in the dataset or not.

Using the Laplace mechanism with these parameters, the credit card company can add noise drawn from a Laplace distribution with a scale parameter of *1/epsilon = 1*. This will add random noise to the computed total in a way that preserves the overall statistical properties of the data but makes it difficult to determine the transaction amounts or times of any individual customer.

For example, the computed total transaction amount might be $10,000, but with the added noise, it might be reported as $10,100. This ensures that the analysis cannot be used to identify the transaction amounts or times of any individual customer with high confidence, while still providing useful information about the overall transaction amounts for the specific time period.

However, suppose the credit card company wants to achieve a higher level of privacy protection and sets the privacy budget to *epsilon = 10* instead of *epsilon = 1*. This means that the added noise will be larger and the analysis will be more private, but it will also be less accurate. For example, the computed total transaction amount might be reported as $15,000 with *epsilon = 10*, which is further from the true value of $10,000.

In summary, differential privacy can be used in the context of fraud detection to protect the privacy of individuals in a dataset while still allowing useful statistical analysis to be performed. However, the choice of privacy budget (epsilon) is important and should be balanced with the level of privacy protection and the desired accuracy of the analysis.

Benefits of differential privacy

Why use differential privacy at all? What benefits will it provide to users? And what benefits, if any, will it provide to engineers, researchers, and scientists? There are several key benefits that differential privacy provides, including strong user privacy guarantees, flexibility in analysis, balance between privacy and utility, robustness, and transparency. These are more important in the cybersecurity domain than others, as we discussed in *Chapter 1, On Cybersecurity and Machine Learning*.

User privacy guarantees

Differential privacy provides a rigorous mathematical definition of privacy protection that offers strong guarantees of privacy. It ensures that an individual's personal data cannot be distinguished from the data of others in the dataset.

In a cybersecurity context, differential privacy can be used to protect the privacy of user data in security logs. For example, let's say a security analyst is examining a log of user login attempts. Differential privacy can be used to protect the privacy of individual users by adding random noise to the log data so that it is impossible to determine whether a specific user attempted to log in.

Flexibility

Differential privacy can be applied to a wide range of data analysis techniques, including queries, machine learning algorithms, and statistical models. In cybersecurity, it can be applied to a variety of security-related data analysis techniques. For example, it can be used to protect the privacy of user data in intrusion detection systems. It can also be applied to the algorithms used by these systems to detect anomalous network activity and to ensure that the privacy of individual users is protected.

Privacy-utility trade-off

Differential privacy provides a way to balance privacy and utility so that accurate statistical analysis can be performed on a dataset while minimizing the risk of exposing sensitive information. It can be used to protect the privacy of sensitive data in cybersecurity applications while still allowing useful insights to be obtained. For example, it can be used to protect the privacy of user data in threat intelligence-sharing systems. It can also be used to protect the privacy of individual users while still allowing organizations to share information about threats and vulnerabilities.

Robustness

Differential privacy is robust to various types of attacks, including statistical attacks and inference attacks. Differential privacy is designed to protect against a wide range of attacks, including statistical attacks and inference attacks. For example, in a cybersecurity context, differential privacy can be used to protect the privacy of user data in forensic investigations. It can also be used to ensure that sensitive data cannot be inferred from forensic evidence, even if an attacker has access to a large amount of other data.

Transparency

Differential privacy provides a way to quantify the amount of privacy protection provided by a particular technique, which allows individuals and organizations to make informed decisions about the level of privacy they need for their data.

It provides a way to measure the effectiveness of privacy protection techniques, which can be useful in making decisions about data protection in cybersecurity. For example, it can be used to protect the privacy of user data in threat modeling. It can also be used to help organizations understand the level of privacy protection they need to protect against various types of threats and to measure the effectiveness of their existing privacy protection measures.

So far, we have looked at privacy and then differential privacy. Now, let us see how it can be practically applied in the context of machine learning.

Differentially private machine learning

In this section, we will look at how a fraud detection model can incorporate differential privacy. We will first look at the library we use to implement differential privacy, followed by how a credit card fraud detection machine learning model can be made differentially private.

IBM Diffprivlib

`Diffprivlib` is an open source Python library that provides a range of differential privacy tools and algorithms for data analysis. The library is designed to help data scientists and developers apply differential privacy techniques to their data in a simple and efficient way.

One of the key features of `Diffprivlib` is its extensive range of differentially private mechanisms. These include mechanisms for adding noise to data, such as the Gaussian, Laplace, and Exponential mechanisms, as well as more advanced mechanisms, such as the hierarchical and subsample mechanisms. The library also includes tools for calculating differential privacy parameters, such as sensitivity and privacy budget (epsilon), and for evaluating the privacy of a given dataset.

Another important feature of `Diffprivlib` is its ease of use. The library provides a simple and intuitive API that allows users to apply differential privacy to their data with just a few lines of code. The API is designed to be compatible with `scikit-learn`, a popular machine learning library for Python, which makes it easy to incorporate differential privacy into existing data analysis workflows.

In addition to its core functionality, `Diffprivlib` includes a number of advanced features and tools that can be used to improve the accuracy and efficiency of differential privacy applications. For example, the library includes tools for generating synthetic datasets that are differentially private, which can be used to test and validate differential privacy mechanisms. It also includes tools for differential private machine learning, which can be used to build models that are both accurate and privacy-preserving.

Overall, `Diffprivlib` provides a powerful set of tools for data privacy that can be used in a wide range of applications, from healthcare and finance to social media and online advertising. Its extensive range of differentially private mechanisms, ease of use, and advanced features make it a valuable resource for anyone looking to improve the privacy and security of their data analysis workflows.

In the following sections, we will use `Diffprivlib` to train and evaluate differentially private machine learning models.

Credit card fraud detection with differential privacy

As we know, differential privacy is a framework for preserving the privacy of individuals while allowing statistical analysis of their data. Many applications today are powered by analysis through machine learning, and hence, the application of DP in machine learning has been a field of growing interest and importance.

To apply differential privacy to a machine learning technique, we will perform the following steps:

1. **Define the privacy budget**: The first step is to define the privacy budget, which determines the level of privacy protection that will be provided. The privacy budget is typically expressed as ε, which is a small positive number. The smaller the value of ε, the stronger the privacy protection, but also the less accurate the results.

2. **Add noise to the data**: To ensure differential privacy, noise is added to the data before logistic regression is performed. Specifically, random noise is added to each data point, so that the noise cancels out when the data is aggregated.

3. **Train the model**: Once the data has been randomized, a machine learning model is trained on the randomized data. This model will be less accurate than a model trained on the original data, but it will still be useful for making predictions.

4. **Evaluate the model**: Once the model has been trained, it can be used to make predictions on new data. The accuracy of the model will depend on the value of ε that was chosen, as well as the size and complexity of the dataset.

In the following sections, we will look at how this can be applied in practice to two popular classification models: logistic regression and random forests.

Differentially private logistic regression

First, we will import the required libraries:

```
from sklearn.model_selection import train_test_split
from sklearn.linear_model import LogisticRegression
from diffprivlib.models import LogisticRegression as
dpLogisticRegression
import pandas as pd
```

As a simulation, we will be using the credit card fraud detection dataset from Kaggle. You can use any dataset of your choice. We split the data into training and test sets, with 2% reserved for testing:

```
import pandas as pd
url = "https://raw.githubusercontent.com/nsethi31/Kaggle-Data-Credit-
Card-Fraud-Detection/master/creditcard.csv"
df = pd.read_csv(url)
```

To print the columns, you can simply run the following:

```
print(df.columns)
```

And you should see the following:

```
Index(['Time', 'V1', 'V2', 'V3', 'V4', 'V5', 'V6', 'V7', 'V8', 'V9', 'V10',
       'V11', 'V12', 'V13', 'V14', 'V15', 'V16', 'V17', 'V18', 'V19', 'V20',
       'V21', 'V22', 'V23', 'V24', 'V25', 'V26', 'V27', 'V28', 'Amount',
       'Class'],
      dtype='object')
```

Figure 10.1 – Dataset columns

We want to use columns V1 through V28 and Amount as features, and Class as the label. We then want to split the data into training and test sets:

```
Y = df['Class'].values
X = df.drop('Time', axis = 1).drop('Class', axis = 1).values

# Split into train and test sets
X_train, X_test, y_train, y_test = train_test_split(X,Y,
test_size=0.2,random_state=123)
```

Next, we train a logistic regression model to predict the class of the data. Note that this is the vanilla logistic regression model from scikit-learn without any differential privacy involved:

```
# Train a regular logistic regression model
model = LogisticRegression()
model.fit(X_train, y_train)
```

Now, we evaluate the performance of this model on the test set:

```
# Evaluate the model on the test set
score = model.score(X_test, y_test)
print("Test set accuracy for regular logistic regression: {:.2f}%".
format(score*100))
```

This should print something like this:

```
Test set accuracy for regular logistic regression: 99.90%
```

Great! We have nearly 99.9% accuracy on the test set.

Now, we fit a differentially private logistic regression model on the same data. Here, we set the value of the `epsilon` parameter to 1. You can set this to any value you want, as long as it is not zero (an epsilon of zero indicates no differential privacy, and the model will be equivalent to the vanilla one):

```
# Train a differentially private logistic regression model
dp_model = dpLogisticRegression(epsilon=1.0, data_norm=10)
dp_model.fit(X_train, y_train)
```

Then, evaluate it on the test set as we did with the previous model:

```
# Evaluate the model on the test set
score = dp_model.score(X_test, y_test)
print("Test set accuracy for differentially private logistic
regression: {:.2f}%".format(score*100))
```

You should see an output like this:

```
Test set accuracy for differentially private logistic regression:
63.73%
```

Wow – that's a huge drop! The accuracy on the test set dropped from 99.9% to about 64%. This is the utility cost associated with increased privacy.

Differentially private random forest

As a fun experiment, let us try the same with a random forest. The code remains almost the same, except both classifiers are switched to random forests. Here's the code snippet:

```
from sklearn.ensemble import RandomForestClassifier
from diffprivlib.models import RandomForestClassifier as
dpRandomForestClassifier

# Train a regular logistic regression model
model = RandomForestClassifier()
model.fit(X_train, y_train)

# Evaluate the model on the test set
score = model.score(X_test, y_test)
print("Test set accuracy for regular RF: {:.2f}%".format(score*100))

# Train a differentially private logistic regression model
dp_model = dpRandomForestClassifier(epsilon=1.0, data_norm=10)
dp_model.fit(X_train, y_train)

# Evaluate the model on the test set
score = dp_model.score(X_test, y_test)
print("Test set accuracy for differentially private RF: {:.2f}%".
format(score*100))
```

This gives the following output:

```
Test set accuracy for regular RF: 99.95%
Test set accuracy for differentially private RF: 99.80%
```

Interestingly, the drop in accuracy in random forests is much less pronounced and is less than 1%. Therefore, random forests would be a better classifier to use in this scenario if both increased privacy and utility are to be achieved.

Examining the effect of ϵ

Now, we will examine how the accuracy of the classifier on the test set varies as we change the value of epsilon. For multiple values of epsilon from 0 to 5, we will train a differentially private classifier and compute the accuracy on the test set:

```
import numpy as np

EPS_MIN = 0.1
EPS_MAX = 10
STEP_SIZE = 0.1
```

```
scores = []

epsilons = np.arange(EPS_MIN, EPS_MAX, STEP_SIZE)

for eps in epsilons:

    # Train a differentially private logistic regression model
    dp_model = dpLogisticRegression(epsilon= eps,data_norm=10)
    dp_model.fit(X_train, y_train)

    # Evaluate the model on the test set
    score = dp_model.score(X_test, y_test)
    scores.append(100.0*score)
```

After this block is run, we can plot the scores against the corresponding `epsilon` values:

```
import matplotlib.pyplot as plt
plt.plot(epsilons, scores)
```

This should show you a plot like this:

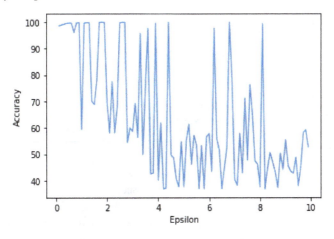

Figure 10.2 – Accuracy variation with epsilon for logistic regression

How about the same evaluation for a random forest? Just replace the model instantiated with a random forest instead of logistic regression. Here is the complete code snippet:

```
import numpy as np

EPS_MIN = 0.1
EPS_MAX = 10
STEP_SIZE = 0.1
```

```
scores = []

epsilons = np.arange(EPS_MIN, EPS_MAX, STEP_SIZE)

for eps in epsilons:

  # Train a differentially private logistic regression model
  dp_model = dpRandomForestClassifier(epsilon= eps,
data_norm=10)
  dp_model.fit(X_train, y_train)

  # Evaluate the model on the test set
  score = dp_model.score(X_test, y_test)
  scores.append(100.0*score)
```

Plotting this gives you the following:

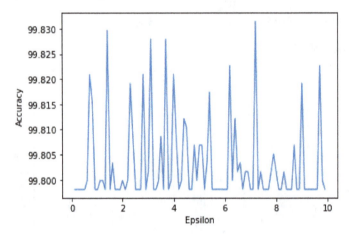

Figure 10.3 – Accuracy variation with epsilon for random forest

The graph *appears* volatile – but note that the accuracy is always between 99.8% and 99.83%. This means that higher values of epsilon do not cause a meaningful difference in accuracy. This model is better suited for differential privacy than the logistic regression model.

Differentially private deep learning

In the sections so far, we covered how differential privacy can be implemented in standard machine learning classifiers. In this section, we will cover how it can be implemented for neural networks.

DP-SGD algorithm

Differentially private stochastic gradient descent (DP-SGD) is a technique used in machine learning to train models on sensitive or private data without revealing the data itself. The technique is based on the concept of differential privacy, which guarantees that an algorithm's output remains largely unchanged, even if an individual's data is added or removed.

DP-SGD is a variation of the **stochastic gradient descent (SGD)** algorithm, which is commonly used for training deep neural networks. In SGD, the algorithm updates the model parameters by computing the gradient of the loss function on a small randomly selected subset (or "batch") of the training data. This is done iteratively until the algorithm converges with a minimum of the loss function.

In DP-SGD, the SGD algorithm is modified to incorporate a privacy mechanism. Specifically, a small amount of random noise is added to the gradients at each iteration, which makes it difficult for an adversary to infer individual data points from the output of the algorithm.

The amount of noise added to the gradients is controlled by a parameter called the privacy budget ε, which determines the maximum amount of information that can be leaked about an individual data point. A smaller value of ε corresponds to a stronger privacy guarantee but also reduces the accuracy of the model.

The amount of noise added to the gradients is calculated using a technique called the Laplace mechanism. The Laplace mechanism adds random noise sampled from the Laplace distribution, which has a probability density function proportional to $exp(-|x|/b)$, where b is the scale parameter. The larger the value of b, the smaller the amount of noise added.

To ensure that the privacy budget ε is not exceeded over the course of the training process, a technique called **moment accountant** is used. Moment accountant estimates the cumulative privacy loss over multiple iterations of the algorithm and ensures that the privacy budget is not exceeded.

DP-SGD differs from a standard SGD only in the gradient calculation step. First, the gradient is calculated for a batch as the partial derivative of the loss with respect to the parameter. Then, the gradients are clipped so that they remain within a fixed window. Finally, random noise is added to the gradients to form the final gradients. This final gradient is used in the parameter update step in gradient descent.

In summary, DP-SGD is a variant of SGD that incorporates a privacy mechanism by adding random noise to the gradients at each iteration. The privacy level is controlled by a parameter called the privacy budget ε, which determines the amount of noise added to the gradients. The Laplace mechanism is used to add the noise, and the moment accountant technique is used to ensure that the privacy budget is not exceeded.

DP-SGD has several advantages over traditional SGD algorithms:

- **Privacy-preserving**: The primary advantage of DP-SGD is that it preserves the privacy of individual data points. This is particularly important when dealing with sensitive or confidential data, such as medical records or financial data.

- **Robustness to re-identification attacks**: DP-SGD provides robustness to re-identification attacks, which attempt to match the output of the algorithm to individual data points. By adding random noise to the gradients, DP-SGD makes it difficult for an attacker to distinguish between individual data points.

- **Improved fairness**: DP-SGD can also improve the fairness of machine learning models by ensuring that the model does not rely too heavily on any individual data point. This can help prevent biases in the model and ensure that it performs well across different demographic groups.

- **Scalability**: DP-SGD can scale to large datasets and complex models. By using SGD, DP-SGD can train models on large datasets by processing small batches of data at a time. This allows for efficient use of computing resources.

- **Accuracy trade-off**: Finally, DP-SGD offers a trade-off between accuracy and privacy. By adjusting the privacy budget ε, the user can control the level of privacy protection while still achieving a reasonable level of accuracy. This makes DP-SGD a flexible and adaptable tool for machine learning applications.

Implementation

We will begin, as usual, by implementing the necessary libraries. Apart from the usual processing and deep learning libraries, we will be using a new one, known as `tensorflow-privacy`. This library provides tools for adding differential privacy to TensorFlow models, including an implementation of the TensorFlow privacy algorithm for training deep learning models with differential privacy. The library also includes tools for measuring the privacy properties of a model, such as its `epsilon` value, which quantifies the level of privacy protection provided by the differential privacy mechanism:

```
import tensorflow as tf
import numpy as np
import tensorflow_privacy
from tensorflow_privacy.privacy.analysis import compute_dp_sgd_privacy
```

We will now write a function that will load and preprocess our MNIST data. The MNIST dataset is a large collection of handwritten digits that is commonly used as a benchmark dataset for testing machine learning algorithms, particularly those related to image recognition and computer vision. The dataset consists of 60,000 training images and 10,000 testing images, with each image being a grayscale 28x28 pixel image of a handwritten digit (0-9).

Our function will first load the training and test sets from this data. The data is then scaled to 1/255th its value, followed by reshaping into the image dimensions. The labels, which are integers from 0 to 9, are converted into one-hot vectors:

```
def load_and_process_MNIST_Data():

    # Define constants
```

```
SCALE_FACTOR = 1/255
NUM_CLASS = 10

# Load train and test data
train, test = tf.keras.datasets.mnist.load_data()
train_data, train_labels = train
test_data, test_labels = test
print("----- Loaded Train and Test Raw Data -----")

# Scale train and test data
train_data = np.array(train_data, dtype=np.float32) * SCALE_FACTOR
test_data = np.array(test_data, dtype=np.float32) * SCALE_FACTOR
print("----- Scaled Train and Test Data -----")

# Reshape data for Convolutional NN
train_data = train_data.reshape(train_data.shape[0], 28, 28, 1)
test_data = test_data.reshape(test_data.shape[0], 28, 28, 1)
print("----- Reshaped Train and Test Data -----")

# Load train and test labels
train_labels = np.array(train_labels, dtype=np.int32)
test_labels = np.array(test_labels, dtype=np.int32)
print("----- Loaded Train and Test Labels -----")

# One-Hot Encode the labels
train_labels = tf.keras.utils.to_categorical(train_labels, num_
classes=NUM_CLASS)
test_labels = tf.keras.utils.to_categorical(test_labels, num_
classes=NUM_CLASS)
print("----- Categorized Train and Test Labels -----")

return train_data, train_labels, test_data, test_labels
```

Next, we will define a function that creates our classification model. In this case, we will be using CNNs. We have seen and used CNNs in earlier chapters; however, we will provide a brief recap here.

A CNN is a type of neural network that is specifically designed for image recognition and computer vision tasks. CNNs are highly effective at processing and analyzing images due to their ability to detect local patterns and features within an image. At a high level, a CNN consists of a series of layers, including convolutional layers, pooling layers, and fully connected layers. In the convolutional layers, the network learns to detect local features and patterns in the input image by applying a set of filters to the image. The pooling layers then downsample the feature maps obtained from the convolutional layers to reduce the size of the input and make the network more computationally efficient. Finally, the fully connected layers process the output of the convolutional and pooling layers to generate a prediction.

The key innovation of CNNs is the use of convolutional layers, which allow the network to learn spatially invariant features from the input image. This is achieved by sharing weights across different parts of the image, which allows the network to detect the same pattern regardless of its position within the image. CNNs have achieved state-of-the-art performance in a wide range of computer vision tasks, including image classification, object detection, and semantic segmentation. They have been used in many real-world applications, such as self-driving cars, medical image analysis, and facial recognition systems.

Our model creation function initializes an empty list and adds layers to it one by one to build up the CNN structure:

```python
def MNIST_CNN_Model (num_hidden = 1):
    model_layers = list()

    # Add input layer

    # Convolution
    model_layers.append(tf.keras.layers.Conv2D(16, 8,
                        strides=2,
                        padding='same',
                        activation='relu',
                        input_shape=(28, 28, 1)))

    # Pooling
    model_layers.append(tf.keras.layers.MaxPool2D(2, 1))

    # Add Hidden Layers
    for _ in range(num_hidden):

        # Convolution
        model_layers.append(tf.keras.layers.Conv2D(32, 4,
                            strides=2,
                            padding='valid',
                            activation='relu'))

        # Pooling
        model_layers.append(tf.keras.layers.MaxPool2D(2, 1))

    # Flatten to vector
    model_layers.append(tf.keras.layers.Flatten())
```

```
# Final Dense Layer
model_layers.append(tf.keras.layers.Dense(32, activation='relu'))
model_layers.append(tf.keras.layers.Dense(10))

# Initialize model with these layers
model = tf.keras.Sequential(model_layers)

return model
```

All of the core functions needed have been defined. Now, we use the data loader we implemented earlier to load the training and test data and labels:

```
train_data, train_labels, test_data, test_labels = load_and_process_
MNIST_Data()
```

Here, we will set some hyperparameters that will be used by the model:

- NUM_EPOCHS: This defines the number of epochs (one epoch is a full pass over the training data) that the model will undergo while training.

- BATCH_SIZE: This defines the number of data instances that will be processed in one batch. Processing here involves running the data through the network, calculating the predicted labels, the loss, and the gradients, and then updating the weights by gradient descent.

- MICRO_BATCHES: The dataset is divided into smaller units called microbatches, with each microbatch containing a single training example by default. This allows us to clip gradients for each individual example, which reduces the negative impact of clipping on the gradient signal and maximizes the model's utility. However, increasing the size of microbatches can decrease computational overhead, but it involves clipping the average gradient across multiple examples. It's important to note that the total number of training examples consumed in a batch remains constant, regardless of the microbatch size. To ensure proper division, the number of microbatches should evenly divide the batch size.

- L2_NORM_CLIP: This refers to the maximum L2-norm that is allowed for the gradient of the loss function with respect to the model parameters. During training, the gradient computed on a minibatch of data is clipped to ensure that its L2-norm does not exceed the L2_NORM_CLIP value. This clipping operation is an essential step in the DP-SGD algorithm because it helps to bind the sensitivity of the gradient with respect to the input data. A higher value can lead to better accuracy but may decrease privacy guarantees, while a lower value can provide stronger privacy guarantees but may result in slower convergence and lower accuracy.

- NOISE_MULTIPLIER: This controls the amount of noise that is added to the gradient updates during training to provide privacy guarantees. In DP-SGD, each gradient update is perturbed by a random noise vector to mask the contribution of individual training examples to the gradient. A higher value increases the amount of noise that is added to the gradient, which in turn provides stronger privacy guarantees but can decrease the accuracy of the model.

- LEARN_RATE: This is the learning rate, and as seen in earlier chapters, controls the degree to which gradients are updated.

Note that the following values we set for these hyperparameters have been derived through experimentation. There is no sure way of knowing what the best parameters are. In fact, you are encouraged to experiment with different values and examine how they affect the privacy and accuracy guarantees of the model:

```
NUM_EPOCHS = 3
BATCH_SIZE = 250
MICRO_BATCHES = 250
L2_NORM_CLIP = 1.5
NOISE_MULTIPLIER = 1.3
LEARN_RATE = 0.2
```

We will initialize the model using the function we defined earlier, and print out a summary to verify the structure:

```
model = MNIST_CNN_Model()
model.summary()
```

This will show you something like this:

```
Model: "sequential_1"

_____
 Layer (type)                 Output Shape              Param #
=================================================================
 conv2d_3 (Conv2D)            (None, 14, 14, 16)        1040

 max_pooling2d_3 (MaxPooling  (None, 13, 13, 16)        0
 2D)

 conv2d_4 (Conv2D)            (None, 5, 5, 32)          8224

 max_pooling2d_4 (MaxPooling  (None, 4, 4, 32)          0
 2D)

 flatten_1 (Flatten)          (None, 512)               0

 dense_2 (Dense)              (None, 32)                16416

 dense_3 (Dense)              (None, 10)                330

=================================================================
Total params: 26,010
Trainable params: 26,010
Non-trainable params: 0
_____
```

Figure 10.4 – Model structure

Now, we will define the loss and optimizer used for training. While the loss is categorical cross-entropy (as expected for a multi-class classification problem), we will not use the standard Adam optimizer here but will use a specialized optimizer for differential privacy.

DPKerasSGDOptimizer is a class in the TensorFlow Privacy library that provides an implementation of the SGD optimizer with differential privacy guarantees. It uses the DP-SGD algorithm, which adds random noise to the gradients computed during each step of the SGD optimization process. The amount of noise added is controlled by two parameters: the noise multiplier and the clipping norm. The noise multiplier determines the amount of noise added to the gradients, while the clipping norm limits the magnitude of the gradients to prevent large updates:

```
optimizer = tensorflow_privacy.DPKerasSGDOptimizer(
            l2_norm_clip = L2_NORM_CLIP,
            noise_multiplier = NOISE_MULTIPLIER,
            num_microbatches = MICRO_BATCHES,
            learning_rate = LEARN_RATE)

loss = tf.keras.losses.CategoricalCrossentropy(
        from_logits=True,
        reduction=tf.losses.Reduction.NONE)
```

Finally, we will build the model and start the training process:

```
model.compile(optimizer=optimizer,
              loss=loss,
              metrics=['accuracy'])

model.fit(train_data,
          train_labels,
          epochs = NUM_EPOCHS,
          validation_data = (test_data, test_labels),
          batch_size = BATCH_SIZE)
```

This should show you the training loop as follows:

```
Train on 60000 samples, validate on 10000 samples
Epoch 1/3
60000/60000 [==============================] - ETA: 0s - loss: 1.1859 - acc: 0.6345 /usr/local/lib/python3.8/dist-packages/keras/engine/training_v1.py
  updates = self.state_updates
60000/60000 [==============================] - 3861s 64ms/sample - loss: 1.1859 - acc: 0.6345 - val_loss: 0.5360 - val_acc: 0.8359
Epoch 2/3
60000/60000 [==============================] - 3813s 64ms/sample - loss: 0.4900 - acc: 0.8640 - val_loss: 0.4220 - val_acc: 0.8924
Epoch 3/3
60000/60000 [==============================] - 3785s 63ms/sample - loss: 0.4313 - acc: 0.8962 - val_loss: 0.3699 - val_acc: 0.9136
<keras.callbacks.History at 0x7f3eed1b14c0>
```

Figure 10.5 – Training loop

The model is now trained. The `compute_dp_sgd_privacy` function is useful for analyzing the privacy properties of a differentially private machine learning model trained using the DP-SGD algorithm. By computing the privacy budget, we can ensure that the model satisfies a desired level of privacy protection and can adjust the parameters of the algorithm accordingly. The function uses the moment accountant method to estimate the privacy budget of the DP-SGD algorithm. This method calculates an upper bound on the privacy budget by analyzing the moments of the privacy loss distribution:

```
compute_dp_sgd_privacy.compute_dp_sgd_privacy(
                    n = train_data.shape[0],
                    batch_size = BATCH_SIZE,
                    noise_multiplier = NOISE_MULTIPLIER,
                    epochs = NUM_EPOCHS,
                    delta=1e-5)
```

And this should show you the following privacy measurements:

```
DP-SGD with sampling rate = 0.417% and noise_multiplier = 1.3 iterated
over 720 steps satisfies differential privacy with eps - 0.563 and
delta = 1e-05.
The optimal RDP order is 18.0.
(0.5631726490328062, 18.0)
```

Differential privacy in practice

Understanding the importance and utility of differential privacy, technology giants have started implementing it in their products. Two popular examples are Apple and Microsoft.

Apple routinely collects users' typing history and behavior locally – this helps power features such as autocorrect and automatic completion of messages. However, it also invites the risk of collecting personal and sensitive information. Users may talk about medical issues, financial details, or other information that they want to protect, and hence using it directly would be a privacy violation. Differential privacy comes to the rescue here. Apple implements **local differential privacy**, which guarantees that it is difficult to determine whether a certain user contributed to the computation of an aggregate feature by adding noise to the data before it is shared with Apple for computation and processing.

Another tech giant that has been a forerunner in differential privacy is Microsoft. The Windows operating system needs to collect telemetry in order to understand usage patterns, diagnose faults, and detect malicious software. Microsoft applies differential privacy to the features it collects by adding noise before they are aggregated and sent to Microsoft. Microsoft Office has a *Suggested Replies* feature, which enables auto-completion and response suggestions in Outlook and Word. As there might be sensitive data in the emails/documents the model is trained on, Microsoft uses differential privacy in order to ensure that the model doesn't learn from or leak any such information.

These algorithms often take longer to train and often require tuning for accuracy, but according to Microsoft, this effort can be worth it due to the more rigorous privacy guarantees that differential privacy enables.

Summary

In recent years, user privacy has grown as a field of importance. Users are to have full control over their data, including its collection, storage, and use. This can be a hindrance to machine learning, especially in the cybersecurity domain, where increased privacy causing a decreased utility can lead to fraud, network attacks, data theft, or abuse.

This chapter first covered the fundamental aspects of privacy – what it entails, why it is important, the legal requirements surrounding it, and how it can be incorporated into practice through the privacy-by-design framework. We then covered differential privacy, a statistical technique to add noise to data so that analysis can be performed while maintaining user privacy. Finally, we looked at how differential privacy can be applied to machine learning in the domain of credit card fraud detection, as well as deep learning models.

This completes our journey into building machine learning solutions for cybersecurity! Now, it is time to introspect and develop more skills in the domain. The next chapter, which contains a series of interview-related questions and additional blueprints, will help you do just that.

11

Protecting User Privacy with Federated Machine Learning

In recent times, the issue of user privacy has gained traction in the information technology world. Privacy means that the user is in complete control of their data – they can choose how the data is collected, stored, and used. Often, this also implies that data cannot be shared with other entities. Apart from this, there may be other reasons why companies may not want to share data, such as confidentiality, lack of trust, and protecting intellectual property. This can be a huge impediment to **machine learning** (**ML**) models; large models, particularly deep neural networks, cannot train properly without adequate data.

In this chapter, we will learn about a privacy-preserving technique for ML known as **federated machine learning** (**FML**). Many kinds of fraud data are sensitive; they have user-specific information and also reveal weaknesses in the company's detection measures. Therefore, companies may not want to share them with one another. FML makes it possible to learn from data without having access to the data by sharing just the learned model itself.

In this chapter, we will cover the following main topics:

- An introduction to federated machine learning
- Implementing federated averaging
- Reviewing the privacy-utility trade-off in **federated learning** (**FL**)

By the end of this chapter, you will have a detailed understanding of federated machine learning, and be able to implement any task (security or non-security related) as a federated learning task.

Technical requirements

You can find the code files for this chapter on GitHub at `https://github.com/PacktPublishing/10-Machine-Learning-Blueprints-You-Should-Know-for-Cybersecurity/tree/main/Chapter%2011`.

An introduction to federated machine learning

Let us first look at what federated learning is and why it is a valuable tool. We will first look at privacy challenges that are faced while applying machine learning, followed by how and why we apply federated learning.

Privacy challenges in machine learning

Traditional ML involves a series of steps that we have discussed multiple times so far: data preprocessing, feature extraction, model training, and tuning the model for best performance. However, this involves the data being exposed to the model and, therefore, is based on the premise of the availability of data. The more data we have available, the more accurate the model will be.

However, there is often a scarcity of data in the real world. Labels are hard to come by, and there is no centrally aggregated data source. Rather, data is collected and processed by multiple entities who may not want to share it.

This is true more often than not in the security space. Because the data involved is sensitive and the stakes are high, entities who collect the data may not want to share it with others or post it publicly.

The following are examples of this:

- Consider a credit card company. The company has labeled data on transactions that were reported to be fraud. However, it may not want to share the data, as exposing the data may unintentionally expose implementation details or intellectual property. User-level information in the data may also be personally identifying, and there may be privacy risks associated with sharing it. Therefore, every credit card company has access to a small set of data – but no company will share it with other companies.

- Consider a network security or antivirus company that uses statistics and ML to detect malware. It will have examples of applications that it has analyzed and manually labeled as malware. However, sharing the data may leak information about attacks and undermine the company's public image. It also may give away clues to attackers on how to circumvent detection. Thus, all antivirus companies will have data but do not share it publicly.

You can draw similar parallels in almost all areas of cybersecurity: click fraud detection, identification of fake news, flagging abusive and hate speech content on social media, and so on.

Entities that own the data are not thrilled about sharing it, which makes training ML models harder. Additionally, no entity can learn from the data acquired by other entities. Valuable knowledge and signals may be lost in the process, and models built on subsets of data will have inherent biases.

Thus, it is extremely challenging, if not impossible, to construct a dataset that is good enough to train an ML model while maintaining privacy. Federated learning is an approach used to overcome this challenge.

How federated machine learning works

FML (or simply FL) is a type of ML that allows multiple devices or organizations to collaborate on building a shared ML model without sharing their data. In traditional ML, all data is aggregated and processed in a central location, but in FML, the data remains distributed across multiple devices or locations, and the model is trained in a decentralized manner.

The fundamental concept in federated learning is to share the learned model instead of sharing the data. Because of this, individual entities can share knowledge about patterns and parameters without having to disclose the data. We will now see how this is achieved in a step-by-step manner.

Step 1 – data partitioning

The data is partitioned across multiple devices or locations, where each one is known as a client. Each client has its own local data that is not shared with other devices. For example, in a medical setting, each hospital may have patient data that it is not willing to share with other hospitals. In the case of fraud detection, each credit card company will have examples of fraud. Note that every client will have data in varying sizes and distributions; however, all clients must preprocess their data in the same way to produce a uniform input vector of data. An example scenario is shown in *Figure 11.1*; there are multiple clients shown, and each client draws data from different sources (a USB drive, a cloud server, a local file, a database, and so on). At each client, however, data is processed into the same standard form:

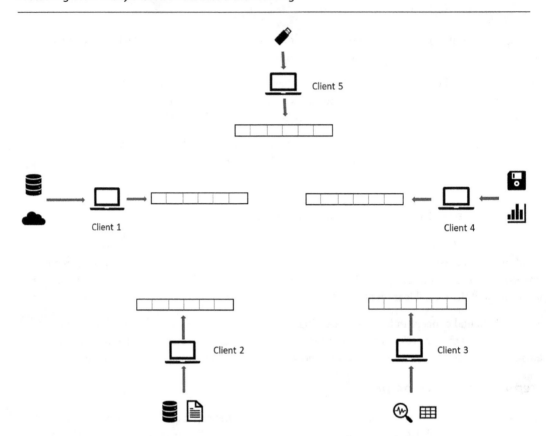

Figure 11.1 – Every client processes the data into a standard form

Step 2 – global model initialization

A central server or client, known as an aggregator, initializes a global model. In this step, the model is initialized randomly, with parameters and weights drawn from a uniform normal distribution. The global model defines the structure and architecture that will be used by each client. The global model is distributed to each client for training, as shown in *Figure 11.2*:

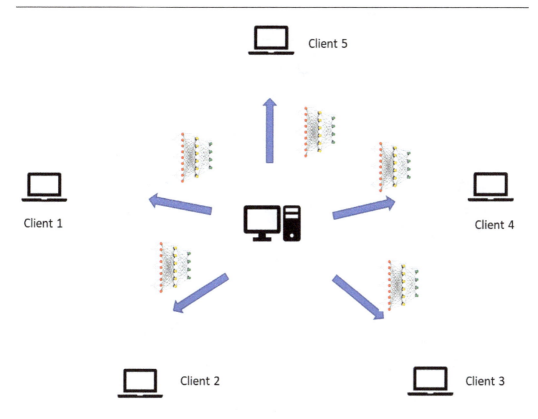

Figure 11.2 – Global model distributed to clients

Step 3 – local training

After receiving the global model, each client trains the local model on the data they have. The local model is trained using standard ML techniques, such as stochastic gradient descent, and updates are made to the local model using the data available on that device. Up until now, the steps we have followed are similar to what an individual client would do in traditional ML. Each client will train and produce its own local model, as shown in *Figure 11.3*:

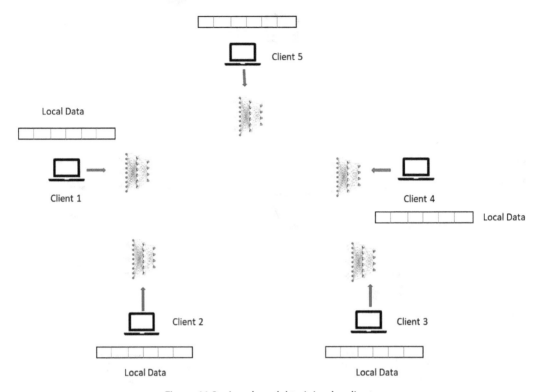

Figure 11.3 – Local model training by clients

Step 4 – model aggregation

The updated local models are then sent back to the central server or cloud, where they are aggregated to create an updated global model. This aggregation process can take different forms, such as averaging the local models or using more complex methods. To calculate the updated aggregated model, the received parameters are averaged and assigned to the new model. Recall that we distribute the global model to each client; thus, each client will have the same structure of the underlying model, which makes aggregation possible. In most cases, the strategy known as **federated averaging** is used. We calculate the average weighted by the size of the data; clients with more data are likely to produce a better model, and so their weights are assigned more importance in the average; this process is demonstrated in *Figure 11.4*:

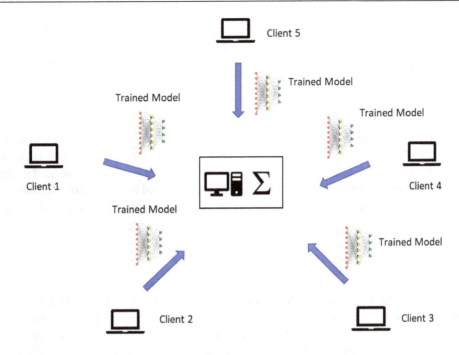

Figure 11.4 – Model aggregation

Step 5 – model distribution

The aggregated model calculated in the previous step is distributed to each client. The client now takes this model and fine-tunes it using the local client data, thus updating it again. This time, however, the initialization is not random. The model parameters are the aggregated ones, meaning that they incorporate the learnings of all client nodes. After updating the model, it is sent back to the central server for aggregation. This continues for a fixed number of communication rounds.

In summary, federated machine learning enables multiple devices or organizations to collaborate on building a shared ML model without sharing their data. This approach can improve data privacy, security, and efficiency, while still providing accurate and useful models.

The benefits of federated learning

In this section, we will discuss the key advantages of federated learning.

Data privacy

Data privacy is one of the most significant advantages of federated ML. In traditional ML approaches, data is collected and stored in a central location, and this can lead to concerns about data privacy and security. In federated learning, however, data remains on local devices or servers, and models are trained without ever sharing the raw data with a central server. This means that data remains secure and there is no risk of a data breach. Additionally, data owners retain control over their data, which is especially important in sensitive industries such as healthcare or finance, where privacy is of utmost importance.

Lower cost

Federated machine learning can be a cost-effective solution for organizations as it allows them to leverage the computational power of multiple devices or servers without having to invest in expensive hardware or cloud computing services. This is because the devices or servers that are used to train the models already exist, and there is no need to purchase additional infrastructure. Additionally, the costs associated with transferring large amounts of data to a central server for analysis are also reduced as data remains on local devices or servers.

Speed

Federated machine learning can also speed up the training process as each device or server can contribute to the training process. This is because models are trained locally on each device or server, and the updates are sent back to the central server, where they are combined to create a global model. Since each device or server is responsible for training only a subset of the data, the training process can be faster than traditional ML approaches. Additionally, because data remains on local devices or servers, there is no need to transfer large amounts of data to a central server, reducing network latency and speeding up the training process.

Performance

FML allows models to be trained on more diverse datasets, leading to improved model accuracy. This is because data is distributed across multiple devices or servers, and each device or server may have slightly different data characteristics. By training models on a diverse set of data, models become more robust and can better generalize to new data. Additionally, models trained using FL can better account for the variability in the data and perform better in real-world scenarios.

Challenges in federated learning

Although federated machine learning offers significant benefits, such as preserving data privacy, there are several challenges that must be addressed to make it practical and effective.

Communication and network latency

Federated machine learning involves multiple parties collaborating to train a model without sharing their data with each other. However, this requires frequent communication and exchange of large amounts of data and model updates between parties. The communication overhead and network latency can be significant, especially when the parties are located in different geographical locations and have limited bandwidth. This can slow down the training process and make it difficult to coordinate the training of the model.

Data heterogeneity

Federated machine learning involves different parties using their own data and hardware to train the model. This can lead to heterogeneity in the data and hardware, making it difficult to design a model that can effectively leverage the strengths of each party while mitigating the weaknesses. For example, some parties may have high-quality data, while others may have noisy data. Some parties may have powerful hardware, while others may have limited processing capabilities. It is important to design a model that can accommodate these differences and ensure that the training process is fair and unbiased.

Data imbalance and distribution shift

In federated learning, the data used by different parties may be highly imbalanced and may have different statistical properties. This can lead to distributional shifts and bias in the model, making it difficult to ensure that the model is fair and unbiased. For example, if one party has significantly more data than the other parties, the model may be biased toward that party's data. It is important to address these issues by carefully selecting the data used for training and by using techniques such as data augmentation and sample weighting to mitigate the effects of data imbalance and distributional shifts.

Free riders and rogue clients

Federated learning is designed to preserve the privacy of the data owned by different parties. However, this also makes it challenging to ensure the security and privacy of the model and the data during training. For example, it may be difficult to ensure that the model updates sent by each party are genuine and have not been tampered with or corrupted. Additionally, it may be difficult to ensure that the data owned by each party remains private and secure, especially when parties are not fully trusted. It is important to develop secure and privacy-preserving techniques for federated learning to mitigate these risks.

Federated optimization

Federated machine learning requires the use of novel optimization techniques that can handle the distributed nature of the training process. These techniques must be able to aggregate the model updates from multiple parties while preserving the privacy and security of the data. This can be challenging, especially when dealing with large-scale datasets. Additionally, the optimization techniques used in federated learning must be efficient and scalable, as the training process can involve a large number of parties and a large amount of data.

Overall, addressing these challenges requires a combination of advanced algorithms, techniques, and infrastructure. Despite these challenges, federated machine learning has the potential to enable a new era of collaborative ML while preserving privacy and security.

This concludes our discussion of federated learning theory. Now that you have a good understanding of how the process works and what the pitfalls involved are, let us implement it practically.

Implementing federated averaging

In this section, we will implement federated averaging with a practical use case in Python. Note that while we are using the MNIST dataset here as an example, this can easily be replicated for any dataset of your choosing.

Importing libraries

We begin by importing the necessary libraries. We will need our standard Python libraries, along with some libraries from Keras, which will allow us to create our deep learning model. The following code snippet imports these libraries:

```
import numpy as np
import random
import cv2
from imutils import paths
import os

# SkLearn Libraries
from sklearn.model_selection import train_test_split
from sklearn.preprocessing import LabelBinarizer
from sklearn.utils import shuffle
from sklearn.metrics import accuracy_score

# TensorFlow Libraries
import tensorflow as tf
from tensorflow.keras.models import Sequential
from tensorflow.keras.layers import Flatten
from tensorflow.keras.layers import Dense
from tensorflow.keras import backend as K
```

Let us now turn to the data we are using.

Dataset setup

The dataset we will be using is the MNIST dataset. The MNIST dataset is a popular benchmark dataset used in ML research. It is a collection of 70,000 grayscale images of handwritten digits, with each image being 28 x 28 pixels in size. The images are split into a training set of 60,000 examples and a test set of 10,000 examples.

The goal of the dataset is to train an ML model to correctly classify each image into the corresponding digit from 0 to 9. The dataset has been widely used for training and testing various ML algorithms such as neural networks, support vector machines, and decision trees. The MNIST dataset has become a standard benchmark for evaluating the performance of image recognition algorithms, and it has been used as a starting point for many new ML researchers. It has also been used in many tutorials and online courses to teach the basics of image recognition and ML:

```python
def load_mnist_data(dataroot):

    X = list()
    y = list()

    for label in os.listdir(dataroot):
        label_dir_path = dataroot + "/"+label

        for imgFile in os.listdir(label_dir_path):
            img_file_path = label_dir_path + "/" + imgFile
            image_gray = cv2.imread(img_file_path, cv2.IMREAD_GRAYSCALE)

            image = np.array(image_gray).flatten()

            X.append(image/255)
            y.append(label)

    return X, y
```

This completes the code to load and preprocess data.

Client setup

Next, we need to write a function that will initialize our client nodes. Note that in the real world, each client or entity will have its own data. However, as we are simulating this scenario, we will implement this manually. The function takes in the data and labels it as input, and returns partitioned data as output. It will break the data into roughly equal chunks and assign each chunk to one client:

```python
def create_client_nodes(X,y,num_clients=10,
prefix='CLIENT_'):

    #create a list of client names
    client_names = []
    for i in range(num_clients):
        client_names.append(prefix + str(i))

    #randomize the data
    data = list(zip(X, y))
    random.shuffle(data)
```

```
#shard data and place at each client
per_client = len(data)//num_clients
client_chunks = []
start = 0
end = 0

for i in range(num_clients):
  end = start + per_client
  if end > len(data):
    client_chunks.append(data[start:])
  else:
    client_chunks.append(data[start:end])
    start = end

  return {client_names[i] : client_chunks[i] for i in
range(len(client_names))}
```

We will also write a helper function that, when given a chunk of data, will shuffle it, prepare it into tensors as needed, and return it to us:

```
def collapse_chunk(chunk, batch_size=32):

    X, y = zip(*chunk)
    dataset = tf.data.Dataset.from_tensor_slices((list(X), list(y)))
    return dataset.shuffle(len(y)).batch(batch_size)
```

It is now time to turn to the modeling.

Model implementation

Now, we will write a function that creates our actual classification model. This function takes in the hidden layer sizes as a parameter in the form of an array. At a high level, this function will perform the following steps:

1. Initialize a sequential Keras model.
2. Create the input layer based on the shape of the input data. Here, it is known to be 784, as we are using the MNIST data.
3. Set the activation function of the input layer to be a **rectified linear unit (ReLU)**.
4. Parse the list of hidden layer sizes passed in as a parameter, and create hidden layers one by one. For example, the [200, 200, 200] parameter means that there are three hidden layers, each with 200 neurons.
5. Create the final layer. This will have the number of nodes equal to the number of classes, which, in this case, is 10.
6. Set the activation of the final layer as softmax so it returns normalized output probabilities.

Here is an implementation of these steps:

```
def MNIST_DeepLearning_Model(hidden_layer_sizes = [200, 200, 200]):
  input_dim = 784
  num_classes = 10

  model = Sequential()

  model.add(Dense(200, input_shape=(input_dim,)))
  model.add(Activation("relu"))

  for hidden in hidden_layer_sizes:
    model.add(Dense(hidden))
    model.add(Activation("relu"))

  model.add(Dense(num_classes))
  model.add(Activation("softmax"))

  return model
```

This function will initialize and return a model of the desired structure.

Weight scaling

Next, we will write a function that scales the weights. Note that as part of federated learning, we will aggregate the weights. If there is a wide disparity between the data sizes, it will reflect in the model performance. A model with smaller training data must contribute less to the aggregate. Therefore, we calculate a scaling factor for each client or node. This factor is simply the proportion of the global training data this client has access to. If there are 1,000 records globally, and client A has 210 records, the scaling factor is 0.21:

```
def scale_weights(all_clients,this_client,weights):

  # First calculate scaling factor

  # Obtain batch size
  batch_size = list(all_clients[this_client])[0][0].shape[0]


# Compute global data size
  sizes = []
  for client in all_clients.keys():
    sizes.append(tf.data.experimental.cardinality(all_
clients[client]).numpy())
```

```
    global_data_size = np.sum(sizes)*batch_size

    # Compute data size in this client
    this_client_size = tf.data.experimental.cardinality(all_
clients[this_client]).numpy()*batch_size

    # Scaling factor is the ratio of the two
    scaling_factor = this_client_size / global_data_size

    scaled_weights = []
    for weight in weights:
      scaled_weights.append(scaling_factor * weight)

    return scaled_weights
```

Global model initialization

Now, let us initialize the global model. This will be the shared central model that will hold the updated parameters in every communication round:

```
global_model = MNIST_DeepLearning_Model(hidden_layer_sizes = [200,
200, 200])
global_model.summary()
```

This will show you the structure of the created model:

```
Model: "sequential_1"

_____
Layer (type)                Output Shape           Param #
================================================================
dense_1 (Dense)             (None, 200)            157000

activation (Activation)     (None, 200)            0

dense_2 (Dense)             (None, 200)            40200

activation_1 (Activation)   (None, 200)            0

dense_3 (Dense)             (None, 200)            40200

activation_2 (Activation)   (None, 200)            0

dense_4 (Dense)             (None, 200)            40200

activation_3 (Activation)   (None, 200)            0

dense_5 (Dense)             (None, 10)             2010

activation_4 (Activation)   (None, 10)             0

================================================================
Total params: 279,610
Trainable params: 279,610
Non-trainable params: 0
_____
```

Figure 11.5 – Model structure

So far, we have written functions to help us with the data and model. Now it is time to put them into action!

Setting up the experiment

We will now use the function we defined earlier to load the training data. After binarizing the labels (that is, converting them into one-hot-encoding form), we will split the data into training and test sets. We will reserve 20% of the data for testing:

```
dataroot = './trainingSet'
X, y = load_mnist_data(dataroot)

y_binarized = LabelBinarizer().fit_transform(y)

X_train, X_test, y_train, y_test = train_test_split(X,
                                    y_binarized,
                                    test_size=0.2,
                                    random_state=123)
```

We will see the function we defined earlier to assign data to clients. This will return us a dictionary of client and data pairs, on which we will apply the chunk-collapsing function we wrote:

```
clients = create_client_nodes(X_train, y_train, num_clients=10)
clients_data = {}
for client_name in clients.keys():
    clients_data[client_name] = collapse_chunk(clients[client_name])

#process and batch the test set
test_batched = tf.data.Dataset.from_tensor_slices((X_test, y_test)).
batch(len(y_test))
```

Next, we will set a few parameters for our experiment. Note that as we have a multi-class classification problem, the choice of categorical cross-entropy as the loss is obvious. The learning rate and the number of communication rounds are hyperparameters – you can experiment with them:

```
learn_rate = 0.01
num_rounds = 40
loss='categorical_crossentropy'
metrics = ['accuracy']
```

Now that we have everything set up, we can actually carry out federated learning.

Putting it all together

Now we will actually carry out the federated learning process in multiple communication rounds. At a high level, here is what we are doing in each communication round:

1. Obtain the weights from the global model and shuffle client chunks.

2. For every client:

    A. Initialize a local model for training.

    B. Set the weights of the local model to the current global model.

    C. Train the local model with data from this client.

    D. Based on the amount of data available in this client, obtain scaled weights from this newly trained model.

3. For memory management purposes, clear the local Keras session.

4. Compute the average of all the weights obtained in *step 2*. As these are scaled weights, the average is a weighted one.

5. Set the average weights to be the weights for the new global model.

6. Evaluate the accuracy of this model (new global model with the newly assigned weights).

7. Print information about loss and accuracy.

8. Repeat *steps 1–7* for each communication round. In every round, the global model is gradually updated.

Here's the code to implement these steps:

```
for round in range(num_rounds):

    # Get the weights of the global model
    global_weights = global_model.get_weights()


    scaled_local_weights = []

    # Shuffle the clients
    # This will remove any inherent bias
    client_names= list(clients_data.keys())
    random.shuffle(client_names)

    # Create initial local models
    for client in client_names:
```

```
        # Create the model
        local_client_model = MNIST_DeepLearning_Model(hidden_layer_
sizes = [200])

        # Compile the model
        local_client_model.compile(loss=loss,
        optimizer=
            tf.keras.optimizers.Adam(learning_rate=0.01),
        metrics=metrics)

        # The model will have random weights
        # We need to reset it to the weights of the current global
model
        local_client_model.set_weights(global_weights)

        # Train local model
        local_client_model.fit(clients_data[client],
epochs=1,verbose = 0)

        # Scale model weights
        # Based on this client model's local weights
        scaled_weights = scale_weights(clients_data, client,local_
client_model.get_weights())

        # Record the value
        scaled_local_weights.append(scaled_weights)

        # Memory management
        K.clear_session()

    # Communication round has ended
    # Need to compute the average gradients from all local models
    average_weights = []
    for gradients in zip(*scaled_local_weights):
        # Calculate mean per-layer weights
        layer_mean = tf.math.reduce_sum(gradients, axis=0)

        # This becomes new weight for that layer
        average_weights.append(layer_mean)

    # Update global model with newly computed gradients
    global_model.set_weights(average_weights)

    # Evaluate performance of model at end of round
    losses = []
    accuracies = []
```

```
    for(X_test, Y_test) in test_batched:
        # Use model for inference
        Y_pred = global_model.predict(X_test)

        # Calculate loss based on actual and predicted value
        loss_fn = tf.keras.losses.CategoricalCrossentropy(from_
logits=True)
        loss_value = loss_fn(Y_test, Y_pred)
        losses.append(loss_value)

        # Calculate accuracy based on actual and predicted value
        accuracy_value = accuracy_score(tf.argmax(Y_pred, axis=1),tf.
argmax(Y_test, axis=1))
        accuracies.append(accuracy_value)



    # Print Information
    print("ROUND: {} ---------- GLOBAL ACCURACY: {:.2%}".format(round,
accuracy_value))
```

The result here will be a series of statements that show how the loss and accuracy changed from round to round. Here is what the first 10 rounds look like:

```
263/263 [==============================] - 1s 5ms/step
ROUND: 0 ---------- GLOBAL ACCURACY: 72.86%
263/263 [==============================] - 1s 3ms/step
ROUND: 1 ---------- GLOBAL ACCURACY: 92.40%
263/263 [==============================] - 1s 2ms/step
ROUND: 2 ---------- GLOBAL ACCURACY: 93.10%
263/263 [==============================] - 1s 2ms/step
ROUND: 3 ---------- GLOBAL ACCURACY: 94.18%
263/263 [==============================] - 1s 2ms/step
ROUND: 4 ---------- GLOBAL ACCURACY: 94.00%
263/263 [==============================] - 1s 2ms/step
ROUND: 5 ---------- GLOBAL ACCURACY: 94.30%
263/263 [==============================] - 1s 3ms/step
ROUND: 6 ---------- GLOBAL ACCURACY: 94.05%
263/263 [==============================] - 1s 2ms/step
ROUND: 7 ---------- GLOBAL ACCURACY: 94.00%
263/263 [==============================] - 1s 2ms/step
ROUND: 8 ---------- GLOBAL ACCURACY: 94.69%
263/263 [==============================] - 1s 2ms/step
ROUND: 9 ---------- GLOBAL ACCURACY: 94.56%
```

Figure 11.6 – First 10 rounds in the training loop

And this is how the last 10 rounds look:

```
ROUND: 40 ---------- GLOBAL ACCURACY: 93.35%
263/263 [==============================] - 1s 3ms/step
ROUND: 41 ---------- GLOBAL ACCURACY: 93.23%
263/263 [==============================] - 1s 2ms/step
ROUND: 42 ---------- GLOBAL ACCURACY: 93.85%
263/263 [==============================] - 1s 2ms/step
ROUND: 43 ---------- GLOBAL ACCURACY: 93.55%
263/263 [==============================] - 1s 2ms/step
ROUND: 44 ---------- GLOBAL ACCURACY: 92.49%
263/263 [==============================] - 1s 2ms/step
ROUND: 45 ---------- GLOBAL ACCURACY: 94.32%
263/263 [==============================] - 1s 2ms/step
ROUND: 46 ---------- GLOBAL ACCURACY: 92.50%
263/263 [==============================] - 1s 2ms/step
ROUND: 47 ---------- GLOBAL ACCURACY: 92.88%
263/263 [==============================] - 1s 2ms/step
ROUND: 48 ---------- GLOBAL ACCURACY: 93.57%
263/263 [==============================] - 1s 3ms/step
ROUND: 49 ---------- GLOBAL ACCURACY: 92.25%
```

Figure 11.7 – Last 10 rounds in the training loop

> **Note**
>
> Due to the random initialization of initial weights and shuffling, the results will change every time we run this. Therefore, the numbers you obtain when you try to recreate this will be different than what is shown here. This is expected.

We can visualize these trends with a plot that will show us the loss over the rounds:

```
import matplotlib.pyplot as plt
plt.plot(range(num_rounds), losses)
plt.xlabel("Communication Rounds")
plt.ylabel("Loss")
```

The output would be something like this:

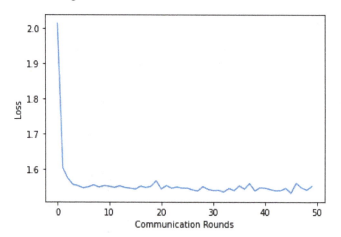

Figure 11.8 – Loss trend over rounds

Similarly, we can plot the accuracy:

```
import matplotlib.pyplot as plt
plt.plot(range(num_rounds), accuracies)
plt.xlabel("Communication Rounds")
plt.ylabel("Accuracy")
```

This will produce the following plot:

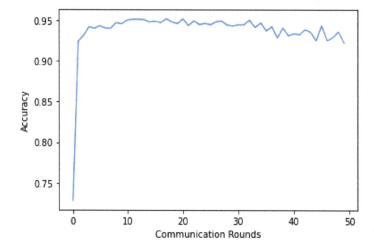

Figure 11.9 – Accuracy trend over rounds

Thus, with federated machine learning, we were able to improve the accuracy of the global model from around 72% to over 92%. Interestingly, you can see that the accuracy somewhat drops around round 25 or 0. This is probably because of overfitting the data on some particular client's local training data.

Reviewing the privacy-utility trade-off in federated learning

In the previous section, we examined the effectiveness of federated learning and looked at the model performance over multiple communication rounds. However, to quantify the effectiveness, we need to compare this against two benchmarks:

- A model trained on the entire data with no federation involved
- A local model trained on its own data only

The differences in accuracy in these three cases (federated, global only, and local only) will indicate the trade-offs we are making and the gains we achieve. In the previous section, we looked at the accuracy we obtain via federated learning. To understand the utility-privacy trade-off, let us discuss two extreme cases – a fully global and a fully local model.

Global model (no privacy)

When we train a global model directly, we use all the data to train a single model. Thus, all parties involved would be publicly sharing their data with each other. The central aggregator would have access to all of the data. In this case, as the data is shared, there is no privacy afforded to clients.

To train the global model on the entire data, we will re-initialize the global model and fit it with the entire training data (instead of individual client data). Here is the code snippet for that:

```
# Initialize global model
global_model = MNIST_DeepLearning_Model(hidden_layer_sizes = [200,
200, 200])
global_model.compile(loss=loss,optimizer=tf.keras.optimizers.
Adam(learning_rate=0.01),metrics=metrics)

# Create dataset from entire data
full_dataset = tf.data.Dataset.from_tensor_slices((X_train, y_
train))\.shuffle(len(y_train))\.batch(32)

# Fit the model
global_model.fit(full_dataset, epochs = 10)
```

You should see output as follows, which shows the accuracy and loss through each epoch:

```
Epoch 1/10
1050/1050 [==============================] - 14s 10ms/step - loss: 0.2968 - accuracy: 0.9119
Epoch 2/10
1050/1050 [==============================] - 8s 7ms/step - loss: 0.1766 - accuracy: 0.9498
Epoch 3/10
1050/1050 [==============================] - 8s 8ms/step - loss: 0.1556 - accuracy: 0.9576
Epoch 4/10
1050/1050 [==============================] - 8s 7ms/step - loss: 0.1342 - accuracy: 0.9632
Epoch 5/10
1050/1050 [==============================] - 7s 6ms/step - loss: 0.1227 - accuracy: 0.9667
Epoch 6/10
1050/1050 [==============================] - 9s 8ms/step - loss: 0.1123 - accuracy: 0.9699
Epoch 7/10
1050/1050 [==============================] - 8s 8ms/step - loss: 0.1153 - accuracy: 0.9710
Epoch 8/10
1050/1050 [==============================] - 8s 7ms/step - loss: 0.1027 - accuracy: 0.9726
Epoch 9/10
1050/1050 [==============================] - 8s 7ms/step - loss: 0.0967 - accuracy: 0.9756
Epoch 10/10
1050/1050 [==============================] - 11s 10ms/step - loss: 0.0894 - accuracy: 0.9776
<keras.callbacks.History at 0x7f790e99f940>
```

Figure 11.10 – Global model training

We see that the fully global model performs much better (it starts at 92% and finishes at 95%) in just a few epochs. This means that it is more powerful than the federated model. This is the trade-off we make in the federated learning; we have to sacrifice performance and accuracy for privacy.

Local model (full privacy)

Now, to evaluate the performance of our local model, we will pick a random client (say client 8) and train a model only on that client's data:

```
# Initialize local client
local_client_model = MNIST_DeepLearning_Model(hidden_layer_sizes =
[200, 200, 200])
local_client_model.compile(loss=loss,optimizer=tf.keras.optimizers.
Adam(learning_rate=0.01),metrics=metrics)

# Train on only one client data
local_client_model.fit(clients_data['CLIENT_8'],epochs=10)
```

The result is as follows:

```
Epoch 1/10
105/105 [==============================] - 5s 12ms/step - loss: 2.1168 - accuracy: 0.1994
Epoch 2/10
105/105 [==============================] - 1s 5ms/step - loss: 1.8923 - accuracy: 0.2699
Epoch 3/10
105/105 [==============================] - 1s 7ms/step - loss: 1.6665 - accuracy: 0.3321
Epoch 4/10
105/105 [==============================] - 1s 7ms/step - loss: 1.5223 - accuracy: 0.4071
Epoch 5/10
105/105 [==============================] - 1s 7ms/step - loss: 1.4313 - accuracy: 0.4167
Epoch 6/10
105/105 [==============================] - 1s 7ms/step - loss: 1.3678 - accuracy: 0.4417
Epoch 7/10
105/105 [==============================] - 1s 5ms/step - loss: 1.3098 - accuracy: 0.4473
Epoch 8/10
105/105 [==============================] - 1s 6ms/step - loss: 1.2857 - accuracy: 0.4539
Epoch 9/10
105/105 [==============================] - 1s 6ms/step - loss: 1.2681 - accuracy: 0.4765
Epoch 10/10
105/105 [==============================] - 1s 10ms/step - loss: 1.2233 - accuracy: 0.4801
```

Figure 11.11 – Local model training

Here, we see that the model started off really poorly (an accuracy of less than 20%, which is worse even than a random guessing classifier). Over the epochs, the model performance improved. However, the accuracy at the end was a little above 48%, which is worse than the federated model.

Understanding the trade-off

The results we have obtained from the previous two subsections show us two things:

- A model that performs federated learning performs better than a model trained only on a client's local data. Thus, a client is able to benefit from the model learned by other clients through parameter sharing and aggregation, which can be achieved without sharing data.

- A global model trained on the entire data performs better than the federated learning model.

This clearly demonstrates the privacy-utility trade-off. As we increase the privacy (apply federated learning), the utility of the data decreases, as reflected in the model performance. At no privacy (all clients share data and use it to train a global model), utility is the highest.

While the trade-off may be harder to understand given the dataset we have used here (after all, what privacy concerns do we face in sharing images of handwritten digits?), it becomes more obvious as we enter the security domain. Recall that in the very first chapter, we discussed how ML in the security domain is different than in other domains, such as image recognition or advertisement targeting, as the stakes are higher here.

Instead of image recognition, consider this to be a credit card fraud detection scenario. The client nodes are various banks and credit card companies. Every company has labeled examples of fraud that they do not want to share.

If the companies share all of the data with each other, it will allow them to train a global model. As different companies will have different patterns and examples of fraud, the model generated will be able to detect multiple attack patterns. This will benefit each company involved; company *A* will be able to detect a certain kind of fraud even if it has never seen it before if another company, *B*, has ever observed it. However, at the same time, companies risk their proprietary data being leaked and personally identifying user data being exposed. There also may be reputational harm if the occurrence of fraud comes to light. Therefore, the high performance and generalizability come at the cost of a loss of privacy.

On the other hand, consider that the companies do not share any data at all. Thus, each company will be able to train a model on the data it has access to. In this case, company *A* will not be able to detect out-of-distribution fraud or novel attack patterns that it has not seen before. This will result in a low recall in detecting fraud and might cause severe losses to the company. Therefore, the high privacy and confidentiality come at the cost of reduced performance and potential financial loss.

Federated learning is able to solve both of these issues and provide us with the perfect middle ground between privacy and utility. Because federated learning involves sharing model parameters and not data, privacy can be maintained, and personal information will not be released. At the same time, because the models are shared and aggregated, learnings from one client will be beneficial to others as well.

Beyond the MNIST dataset

In this chapter, we chose the MNIST dataset for our experiments. While MNIST is a popular benchmark for image processing, it is not ideal for security applications. However, we chose it for two reasons. First, it is a fairly large and well-distributed dataset that makes federated learning simpler. Other public datasets are relatively small, which means that when sharded, clients have only a small amount of data that is not enough to produce a reasonable model. Second, being an image dataset, it is naturally suited to be processed by neural networks.

The applications of federated learning are not limited to images. You are encouraged to explore all of the problems we have looked at in this book so far (malware, fake news, and intrusion detection) and implement them in a federated manner. To do so, only two changes are required:

- Change the data loading mechanism to read from the appropriate data source instead of MNIST
- Change the deep learning model structure to reflect the input and output dimensions specific to the dataset

The rest of the steps (initializing the model, sharding the data, communication rounds, and so on) all remain the same. You should implement federated learning in various cybersecurity areas and observe the privacy-utility trade-off.

Summary

In this chapter, we learned about a privacy preservation mechanism for ML known as federated learning. In traditional ML, all data is aggregated and processed in a central location, but in FML, the data remains distributed across multiple devices or locations, and the model is trained in a decentralized manner. In FML, we share the model and not the data.

We discussed the core concepts and working of FML, followed by an implementation in Python. We also benchmarked the performance of federated learning against traditional ML approaches to examine the privacy-utility trade-off. This chapter provided an introduction to an important aspect of ML and one that is gaining rapid traction in today's privacy-centric technology world.

In the next chapter, we will go a step further and look at the hottest topic in ML privacy today – differential privacy.

12

Breaking into the Sec-ML Industry

This book has covered a broad and diverse set of problems in cybersecurity, along with novel and advanced machine learning solutions to tackle them. Security problems arise in every industry, be it social media, marketing, or information technology. While machine learning for cybersecurity is a hot topic, there are very few resources on how to break into this space. This final chapter covers how to do just that. First, we will look at a set of online resources that you can use to further your understanding of machine learning, cybersecurity, and their intersection. We will also look at a few interview questions that will test your knowledge and help you prepare for interviews. Finally, we will conclude by providing some additional project ideas that you can explore to build your portfolio.

In this chapter, we will cover the following main topics:

- A study guide for machine learning and cybersecurity
- Interview questions
- Additional project blueprints

By the end of this chapter, you will be armed with a set of resources to use to practice and fine-tune your skills as well as anticipate various question patterns.

Study guide for machine learning and cybersecurity

In this section, we will cover some of the resources that can be used to understand machine learning and cybersecurity beyond what we have covered, expanding your knowledge in these areas.

Machine learning theory

Here are a few resources where you can study data science and machine learning:

- **Andrew Ng's YouTube channel** (available as a playlist of videos on YouTube: `https://www.youtube.com/playlist?list=PLLssT5z_DsK-h9vYZkQkYNWcItqhlRJLN`): Andrew Ng is a professor of computer science at Stanford University. His machine learning course (made originally for Coursera) is world-famous. This course explains machine learning from the very basics, in clear and concise terms. You will learn about linear and logistic regression, gradient descent, and neural networks. The course explains the basics as well as the math behind it, with simple examples. All of the exercises in the course are in Matlab; however, you can try and implement them in Python for additional practice. (There is a Python version of the course as well, but I highly recommend that you do the Matlab version and implement it in Python by yourself, which will strengthen your Python coding skills.)

- *Data Mining Concepts and Techniques*: This book by Jiawei Han, Jian Pei, and Micheline Kamber is the flagship book for studying data mining fundamentals. While Andrew Ng's course focuses more on the math behind machine learning, this book focuses on the applications and usefulness of data mining techniques for large datasets. It covers all the phases of machine learning – from preprocessing data to tuning models. It serves as a good reference for understanding common models and algorithms without going into the nitty-gritty mathematical details.

- *Elements of Statistical Learning*: This book by Trevor Hastie, Robert Tibshirani, and Jerome Friedman is a good one for those who want to understand machine learning and the mathematical foundations behind it. The book covers a broad range of topics (supervised and unsupervised learning, kernel smoothing, ensemble models, model selection strategies, and graphical models). The emphasis in this book is more on the statistical and mathematical foundations of machine learning rather than the applications.

- *Deep Learning*: This book by Ian Goodfellow (known as the father of deep learning) is the bible for deep learning and neural networks. This is a book for advanced readers and not beginners. It covers the mathematical foundations of deep learning (linear algebra, probability theory, and numerical methods) as well as deep learning research (RNN, CNN, LSTM, and autoencoders). The book also contains detailed explanations behind the regularization and optimization of deep learning methods.

Hands-on machine learning

The books we discussed so far are meant for those of you who are interested in the theory behind machine learning, and the math behind the operations. However, they offer little guidance on actual, real-world implementation. If you want to learn how to implement these methods as you learn, here are a few books that can help you:

- *Machine Learning Security Principles* by John Paul Mueller

- *Hands-On Graph Neural Networks Using Python* by Maxime Labonne

- *Machine Learning with PyTorch and Scikit-Learn* by Sebastian Raschka, Yuxi (Hayden) Liu, Vahid Mirjalili, and Dmytro Dzhulgakov

Cybersecurity

For data scientists working in the cybersecurity space, a basic understanding of security is generally required. Depending on the industry, you may need to know about cryptography, blockchain, network security, malware, or socio-technical aspects of cybersecurity. Generally, most textbooks on security focus on core security concepts, such as encryption, signature schemes, and hashing. While these are important concepts, they are not strictly necessary for data scientists.

Here are some books that are useful for learning about cybersecurity concepts. You may choose to read those that most closely align with your interests and field:

- *Wireless Communications and Networks*: This book by William Stallings is considered to be the master reference for computer network security. In this book, you will learn about the fundamentals of networking, how the internet works, how wireless communication such as Bluetooth works, what the design principles are, what the security flaws are, and potential flaw-detection mechanisms. The book is helpful if your work involves intrusion detection, log file analysis, or the **Internet of Things (IoT)**.

- *Social Media Security*: This book by Michael Cross provides an in-depth analysis of social networks, including their fundamentals as well as potential security issues. The book covers the dark side of social media, including hacking, social engineering, and digital forensics. This book is a must-read if your work concerns social media platforms.

- *Practical Malware Analysis – The Hands-On Guide to Dissecting Malicious Software*: This book by Michael Sikorski presents an end-to-end guide on malware detection. You will learn how to set up virtual environments, extract network signatures, read malware code, and analyze malware using the latest software. This book is helpful if your work as a data scientist involves malicious applications (ransomware, malware, virus, and harmful web extensions).

Interview questions

In this section, we will look at a few questions that may be asked in data scientist interviews, with an emphasis on those related to cybersecurity or associated topics. We will leave finding the right answers an exercise for you.

Theory-based questions

These questions are theoretical and used to test your knowledge and understanding of machine learning principles and concepts.

Fundamental concepts

- How is machine learning different from traditional computing?

- What is the difference between supervised and unsupervised learning?

- What is semi-supervised learning? Give an example scenario where semi-supervised learning would be the appropriate choice for modeling a problem.

- What is self-supervised learning?

- What is overfitting and how can it be prevented?

- What is underfitting and how can it be prevented?

- How can you detect overfitting/underfitting in a model?

- How does gradient descent work? What is the difference between **SGD** and **Batch-SGD** algorithms?

Data preprocessing

- How are missing values generally handled in datasets?

- Why are missing values present in data? Give some real-world examples where missing values occur.

- In what circumstances is dropping an entire record with missing data appropriate? In what circumstances is dropping an entire column from all records appropriate?

- How can continuous values be converted into categorical features?

- Why does noise occur in a dataset?

- How can the number of features be reduced to improve efficiency?

- How can noise or outliers affect data modeling?

Models

- How does a decision tree find the best split point for features?

- How are split points for categorical features calculated?

- What is the difference between the Gini index and information gain?

- Why do we expect random forests to show improved accuracy compared to decision trees?

- How do random forests work?

- How is linear regression different from logistic regression?

- How does an SVM work?

- What are the advantages of using z-score for anomaly detection? What are the disadvantages?

- What are activation functions? How do you decide which activation function to use?

- How does backpropagation work?

- Neural networks are considered to be a *black box*. Do you agree or disagree? Why?

- In a neural network, how do you determine the following?

  - The number of neurons in the input layer

  - The number of hidden layers

  - The number of neurons in each hidden layer

  - The number of neurons in the output layer

- What is regularization and how can it help?

- How can you prevent a neural network from overfitting?

- What are the advantages (and drawbacks) of the **Local Outlier Factor** (**LOF**) over the z-score?

Text processing

- What is the difference between embeddings generated by **Word2Vec** and those generated by **BERT**?

- How can you calculate the embedding for a sentence if you are given a model such as **Word2Vec** to obtain the embedding of a particular word?

- How is text preprocessed?

- What is TF-IDF? What does a word having a high TF-IDF value indicate?

- What are transformers and how do they work?

- What are attention and self-attention?

- What advantage does attention provide over the LSTM network?

Model tuning and evaluation

- What is more important – precision or recall? Does it vary on a case-by-case basis?

- Why is accuracy not a sufficient metric for evaluating a model? Why do we need precision and recall?

- Why is there a trade-off between precision and recall? How can the best operating point be chosen to maximize both?

- What is cross-validation and why is it used?

- What is stratified cross-validation?

- How do you select the appropriate value of K in K-fold cross-validation?

- What is the F1 score, and why is it useful?

Privacy

- How is security different from privacy?

- What is privacy by design?

- How do you sanitize datasets to ensure that privacy is preserved?

- What is the privacy-utility trade-off? How can it be handled?

- What are the privacy considerations you need to take into account while building an end-to-end machine learning system?

- How does privacy affect machine learning?

- In your opinion, has the GDPR affected the way machine learning is applied?

Experience-based questions

- Can you give an example of a dataset you had to clean for analysis in the cybersecurity field? What types of issues did you encounter, and how did you address them?

- Can you describe a time when you had to deal with missing data in a cybersecurity dataset? How did you impute the missing values, and what impact did this have on your analysis?

- How do you ensure that sensitive information is not included in your data when cleaning it for analysis in the cybersecurity field? Can you give an example of how you handled this in the past?

- Think back to a time when you had to model some data but the labels available to you were very noisy. How did you handle this situation?

- Describe a time when you had to sacrifice recall for precision. How did you make that call, and what were your considerations in the trade-off?

- Can you describe a time when you had to train a machine learning model but the data contained sensitive attributes that, if excluded, showed a degraded model performance? How did you handle this?

Conceptual questions

- When cleaning data to prepare it for machine learning, how do you distinguish between *noise* (which is incorrect or erroneous data) and *anomalies* (which are deviations from normal behavior)?

- A credit card company uses a customer's age as a feature in its machine learning models. To detect noise in the *age* feature, what techniques can be used? Can z-score be used? Why or why not?

- A government agency has a dataset containing information about known terrorists, including their nationality and the types of attacks they have carried out. What techniques might you use to handle missing data in this dataset? Can you provide an example of how this might work in practice?

- A financial institution has a dataset containing information about customer transactions, including the transaction amount and the location where the transaction took place. They notice that the dataset contains many outliers. What techniques might you use to identify and remove these outliers? Can you provide an example of how this might work in practice?

- An e-commerce company wishes to build a model to detect and ban abusive users from its platform. Between decision trees, logistic regression, and a neural network, answer the following questions:

  - Which is the best choice and why?

  - Which is the worst choice and why?

  - What factors did you take into consideration while answering the preceding two questions?

- A malware detector aims to classify malware as one of eight known families. While this is a classification problem, can it be modeled as a regression problem to predict the class from one to eight? Why or why not?

- You are asked to build an anomaly detection model using system logs. You have access to all of the system logs (they are stored indefinitely) but no ground truth or labels:

  - How do you model this data to detect anomalies or attacks without any prior labels?

  - What assumptions do you need to validate for your model to work?

  - What data can you extract from the logs to build your model?

- In a malware detector, you want to identify anomalous API call sequences. Given a large collection of only benign applications (no malware), how can you build a model to detect anomalous API call sequences?

- A hate speech classifier is to be built and the data available for training contains emojis. How can you handle these emojis in your classification pipeline?

- You work for a social media company and are responsible for developing models to detect fake accounts:

  - What data/features will you collect?

  - How will you generate ground truth or labels?

  - How often will you retrain the model?

- A social media company wants to build models for hate speech detection. To do this, several user features need to be collected. How can you maintain privacy while extracting features for the following:

  - Location

- Age
- Private messages

- Consider that you have to train a model to detect child pornography on a video streaming platform:

  - You may need child pornography data to train a model. What are the legal consequences of having such data?

  - How can you train a model without having access to such data?

Additional project blueprints

So far, we have looked at a variety of interesting problems in cybersecurity and explored machine learning solutions for them. However, to really learn and excel in the field, you need to explore and build projects on your own. This section will provide you with blueprints for additional projects. By completing them, you will definitely improve your résumé.

Improved intrusion detection

Cybersecurity has become a critical aspect of our digital world, and machine learning plays an increasingly important role in cybersecurity. ML can help detect and prevent cyberattacks by learning from past incidents and identifying patterns in data. However, the integration of ML into cybersecurity also raises new challenges and potential vulnerabilities, such as adversarial attacks, data poisoning, and model interpretability.

One potential research project on the intersection of cybersecurity and ML is to develop a robust and effective ML-based system for intrusion detection. Intrusion detection is the process of identifying malicious activities and attacks on a computer network. The core problem is building an ML model that can accurately classify network traffic as either normal or malicious.

Several publicly available datasets can be used for this research project, including the following:

- **NSL-KDD**: This dataset is a modified version of the KDDCUP99 dataset, which is commonly used to evaluate intrusion detection systems. The NSL-KDD dataset contains a set of preprocessed features and labels, making it suitable for training ML models.

- **CICIDS2017**: This dataset contains network traffic captured in a research lab environment, with various types of attacks injected into the network. The dataset includes both raw and preprocessed features, as well as ground truth labels.

Several ML techniques can be used to build an intrusion detection system, including the following:

- **Supervised learning**: Training a classification model on labeled data to distinguish between normal and malicious traffic. Common algorithms include logistic regression, decision trees, random forests, and support vector machines.

- **Unsupervised learning**: Using clustering or anomaly detection algorithms to identify patterns in network traffic that deviate from normal behavior.

- **Deep learning**: Training a neural network to learn complex representations of network traffic and detect malicious behavior.

To improve the robustness and reliability of an ML-based intrusion detection system, several techniques can be applied, such as the following:

- **Adversarial training**: Training a model with adversarial examples to improve its resilience to attacks

- **Federated learning**: Collaboratively training a model on multiple devices or networks to protect against data privacy concerns

- **Explainability**: Using interpretability techniques to understand how a model makes predictions and identify potential vulnerabilities or biases

Overall, this research project has the potential to contribute to the development of more effective and secure ML-based intrusion detection systems, which are essential to protect against cyberattacks.

Adversarial attacks on intrusion detection

With the growing sophistication of cyberattacks, traditional security measures such as firewalls and intrusion detection systems are no longer enough to protect organizations from cyber threats. ML algorithms have shown great potential in detecting and mitigating cyberattacks, becoming increasingly popular in cybersecurity research. However, ML models can also be vulnerable to adversarial attacks, where attackers can manipulate input data to mislead the model and compromise its security. Therefore, there is a need to develop robust and secure ML-based cybersecurity solutions.

The core problem of this project is to investigate the vulnerability of ML-based cybersecurity solutions to adversarial attacks and develop robust defense mechanisms against them. Specifically, the project can focus on the following research questions:

- How can ML models be attacked by adversaries in the context of cybersecurity?

- What are the potential consequences of such attacks on the security of a system?

- How can we develop robust ML-based cybersecurity solutions that can defend against adversarial attacks?

There are several publicly available datasets that can be used for this project, such as the following:

- **NSL-KDD**: A benchmark dataset for **intrusion detection systems (IDSs)**, which contains a large number of network traffic records with known *attack* and *normal* labels

- **CICIDS2017**: Another dataset for IDSs, which contains both benign and malicious traffic in a real-world network environment

- **UNB ISCX IDS 2012**: A dataset of network traffic for IDS evaluation, which contains various types of attacks in a controlled environment

To investigate the vulnerability of ML models to adversarial attacks, the project can use various methods, such as the following:

- **Adversarial training**: Training ML models, using adversarial examples to improve their robustness to attacks

- **Adversarial detection**: Developing methods to detect whether an input is adversarial or not, to avoid making decisions based on malicious data

- **Model-agnostic attacks:** Using attacks that are not specific to a particular ML model but, instead, exploit vulnerabilities in the ML pipeline (for example, feature engineering and data preprocessing)

- **Model-specific attacks**: Developing attacks that are tailored to a particular ML model to exploit its weaknesses

To develop robust ML-based cybersecurity solutions, the project can use various techniques, such as the following:

- **Feature selection**: Selecting relevant features for an ML model to improve its performance and reduce its vulnerability to attacks

- **Model ensembling**: Combining multiple ML models to improve their accuracy and robustness to attacks

- **Explainable AI**: Developing ML models that are transparent and explainable, enabling better understanding and interpretation of their decisions

Overall, this project has the potential to make significant contributions to the field of cybersecurity by addressing the critical challenge of developing robust and secure ML-based cybersecurity solutions.

Hate speech and toxicity detection

Hate speech, toxicity, and abuse are prevalent issues on online platforms, and they can have significant negative impacts on individuals and communities. To combat this problem, automated systems that can detect and moderate hate speech and abusive content are needed. ML algorithms have shown promise in this domain, and there is a growing interest in developing robust and effective ML-based hate speech and abuse detection systems.

The core problem of this project is to investigate the effectiveness of ML-based hate speech and abuse detection systems, developing more accurate and robust models. Specifically, the project can focus on the following research questions:

- How can ML models be trained to detect hate speech and abusive content?

- What are the limitations and challenges of existing ML-based hate speech and abuse detection systems?

- How can we develop more robust and effective ML-based hate speech and abuse detection systems?

There are several publicly available datasets that can be used for this project, such as the following:

- **Twitter Hate Speech**: A dataset of tweets labeled as hate speech or not

- **Wikipedia Talk Labels**: A dataset of comments from Wikipedia labeled as toxic or not

- **Hateful Memes Challenge**: A dataset of memes labeled as hateful or not

To investigate the effectiveness of ML-based hate speech and abuse detection systems, the project can use various methods, such as the following:

- **Feature selection**: Identifying relevant features that can capture the characteristics of hate speech and abusive content, such as specific keywords, syntactic patterns, or semantic relationships

- **Model selection**: Evaluating different types of ML models, such as deep learning models, support vector machines, or decision trees, to determine which ones perform better in detecting hate speech and abusive content

- **Data augmentation**: Generating synthetic data to improve the diversity and quality of the training data, and to prevent overfitting

- **Ensemble methods**: Combining multiple ML models to improve the overall performance and robustness of the hate speech and abuse detection system

To develop more accurate and robust ML-based hate speech and abuse detection systems, the project can use various techniques, such as the following:

- **Active learning**: Selecting the most informative samples to label by leveraging human-in-the-loop feedback to improve the accuracy of the model

- **Explainable AI**: Developing ML models that are transparent and explainable, enabling better understanding and interpretation of their decisions

- **Multimodal analysis**: Incorporating additional modalities such as images or audio to improve the accuracy of the hate speech and abuse detection system

Overall, this project has the potential to make significant contributions to the field of hate speech and abuse detection by developing more accurate and robust ML-based systems, helping to combat hate speech and toxicity on online platforms.

Detecting fake news and misinformation

Fake news and misinformation have become critical problems in the modern information age, where the widespread use of social media platforms has enabled the rapid spread of false information. To address this problem, automated systems that can detect and filter out fake news and misinformation are needed. ML algorithms have shown promise in this domain, and there is a growing interest in developing accurate and reliable ML-based fake news and misinformation detection systems.

The core problem of this project is investigating the effectiveness of ML-based fake news and misinformation detection systems and developing more accurate and reliable models. Specifically, the project can focus on the following research questions:

- How can ML models be trained to detect fake news and misinformation?

- What are the limitations and challenges of existing ML-based fake news and misinformation detection systems?

- How can we develop more accurate and reliable ML-based fake news and misinformation detection systems?

There are several publicly available datasets that can be used for this project, such as the following:

- **Fake News Challenge**: A dataset of news articles labeled as fake or real

- **LIAR**: A dataset of statements labeled as true, false, or misleading

- **Clickbait Challenge**: A dataset of headlines labeled as clickbait or not

To investigate the effectiveness of ML-based fake news and misinformation detection systems, the project can use various methods, such as the following:

- **Text analysis**: Analyzing the linguistic and stylistic features of the text, such as sentence structure, sentiment, and readability, to identify patterns that distinguish fake news from real news

- **Network analysis**: Examining the social network structure and dynamics, such as the propagation and diffusion of news articles, to detect patterns of misinformation and disinformation

- **Source credibility analysis**: Assessing the credibility and reliability of the news sources, such as their history, reputation, and affiliations, to evaluate the trustworthiness of the news articles

To develop more accurate and reliable ML-based fake news and misinformation detection systems, the project can use various techniques, such as the following:

- **Ensemble methods**: Combining multiple ML models, such as decision trees, support vector machines, or deep learning models, to improve the overall performance and robustness of a fake news and misinformation detection system

- **Transfer learning**: Adapting pre-trained ML models, such as language models or neural networks, to the specific task of fake news and misinformation detection, leveraging the knowledge learned from other domains.

- **Explainable AI**: Developing ML models that are transparent and explainable, enabling a better understanding and interpretation of their decisions.

Overall, this project has the potential to make significant contributions to the field of fake news and misinformation detection by developing more accurate and reliable ML-based systems, helping to combat the spread of false information and promote fact-based journalism.

Summary

This chapter provided a comprehensive guide to breaking into the Sec-ML industry. It contains all the tools, tricks, and tips that you need to become a data scientist or ML practitioner in the domain of cybersecurity. We began by looking at a set of resources that you can leverage to study ML – both conceptually and hands-on. We also provided several references to books that will help you with the hands-on implementation of ML models in security-related fields. We also shared a question bank that contains commonly asked theory questions in data science interviews, followed by some conceptual, case study-based questions. While neither the resources nor the interview questions are exhaustive, they provide a good starting point.

Finally, the skills and knowledge you have learned so far in this book are of no use if you do not apply them to boost your portfolio. To facilitate this, four project blueprints were presented, along with helpful hints on implementation. We encourage you to follow through on these projects, build them on your own, and add them to your résumé.

We have come to the end of our journey! We began with the basics, learning about the fundamental concepts of ML and cybersecurity, and how the two come together. We looked at multiple problems in cybersecurity (such as malware detection, authorship obfuscation, fake news detection, and automated text detection) and applied novel methods to solve them (transformers, graph neural networks, and adversarial ML). We also looked at a closely related area, privacy, and covered how privacy-preserving mechanisms can be used in ML. Finally, this chapter tied everything together with interview questions and further project ideas.

I hope that this journey was an enlightening one and that you were able to gain a lot of skills along the way. I'm sure that you are now well prepared to take on data science and ML roles in the security field!

Index

www.packtpub.com

Subscribe to our online digital library for full access to over 7,000 books and videos, as well as industry leading tools to help you plan your personal development and advance your career. For more information, please visit our website.

Why subscribe?

- Spend less time learning and more time coding with practical eBooks and Videos from over 4,000 industry professionals

- Improve your learning with Skill Plans built especially for you

- Get a free eBook or video every month

- Fully searchable for easy access to vital information

- Copy and paste, print, and bookmark content

Did you know that Packt offers eBook versions of every book published, with PDF and ePub files available? You can upgrade to the eBook version at packtpub.com and as a print book customer, you are entitled to a discount on the eBook copy. Get in touch with us at customercare@packtpub.com for more details.

At www.packtpub.com, you can also read a collection of free technical articles, sign up for a range of free newsletters, and receive exclusive discounts and offers on Packt books and eBooks.

Other Books You May Enjoy

If you enjoyed this book, you may be interested in these other books by Packt:

Machine Learning Security Principles

John Paul Mueller

ISBN: 978-1-80461-885-1

- Explore methods to detect and prevent illegal access to your system
- Implement detection techniques when access does occur
- Employ machine learning techniques to determine motivations
- Mitigate hacker access once security is breached
- Perform statistical measurement and behavior analysis
- Repair damage to your data and applications
- Use ethical data collection methods to reduce security risks

Hands-On Graph Neural Networks Using Python

Maxime Labonne

ISBN: 978-1-80461-752-6

- Understand the fundamental concepts of graph neural networks
- Implement graph neural networks using Python and PyTorch Geometric
- Classify nodes, graphs, and edges using millions of samples
- Predict and generate realistic graph topologies
- Combine heterogeneous sources to improve performance
- Forecast future events using topological information
- Apply graph neural networks to solve real-world problems

Packt is searching for authors like you

If you're interested in becoming an author for Packt, please visit `authors.packtpub.com` and apply today. We have worked with thousands of developers and tech professionals, just like you, to help them share their insight with the global tech community. You can make a general application, apply for a specific hot topic that we are recruiting an author for, or submit your own idea.

Share Your Thoughts

Now you've finished *10 Machine Learning Blueprints You Should Know for Cybersecurity*, we'd love to hear your thoughts! Scan the QR code below to go straight to the Amazon review page for this book and share your feedback or leave a review on the site that you purchased it from.

https://packt.link/r/1-804-61947-7

Your review is important to us and the tech community and will help us make sure we're delivering excellent quality content.

Download a free PDF copy of this book

Thanks for purchasing this book!

Do you like to read on the go but are unable to carry your print books everywhere? Is your eBook purchase not compatible with the device of your choice?

Don't worry, now with every Packt book you get a DRM-free PDF version of that book at no cost.

Read anywhere, any place, on any device. Search, copy, and paste code from your favorite technical books directly into your application.

The perks don't stop there, you can get exclusive access to discounts, newsletters, and great free content in your inbox daily

Follow these simple steps to get the benefits:

1. Scan the QR code or visit the link below

https://packt.link/free-ebook/9781804619476

2. Submit your proof of purchase
3. That's it! We'll send your free PDF and other benefits to your email directly